# GLOBAL ENVIRONMENTAL CHANGE

## *ITS NATURE AND IMPACT*

**JOHN J. HIDORE**

Prentice Hall
Upper Saddle River, NJ 07458

**Library of Congress Cataloging-in-Publication Data**

Hidore, John J.
Global environmental change : its nature and impact / John J. Hidore.
p. cm.
Includes bibliographical references and index.
ISBN 0–02–354134–2
1. Climatic changes. 2. Biological diversity. 3. Man—Influence on nature. I. Title.
QC981.8.C5H53 1996
551.6—dc20 95–46750
CIP

Acquisitions Editor: Robert McConnin
Editorial Assistant: Grace Anspake
Production Editor: Rose Kernan
Production Coordinator: Trudy Pisciotti
Marketing Manager: Leslie Cavaliere
Marketing Assistant: Amy Reed
Cover Director: Jayne Conte
Cover Art: Jeffrey Patton

Simon & Schuster/A Viacom Company
Upper Saddle River, NJ 07458



Printed in the United States of America

10 9 8 7 6 5 4 3 2 1

ISBN 0-02-354134-2

PRENTICE-HALL INTERNATIONAL (UK) LIMITED, *London*
PRENTICE-HALL OF AUSTRALIA PTY. LIMITED, *Sydney*
PRENTICE-HALL CANADA INC., *Toronto*
PRENTICE-HALL HISPANOAMERICANA, S.A., *Mexico*
PRENTICE-HALL OF INDIA PRIVATE LIMITED, *New Delhi*
PRENTICE-HALL OF JAPAN, INC., *Tokyo*
SIMON & SCHUSTER ASIA PTE. LTD., *Singapore*
EDITORA PRENTICE-HALL DO BRASIL, LTDA., *Rio de Janeiro*

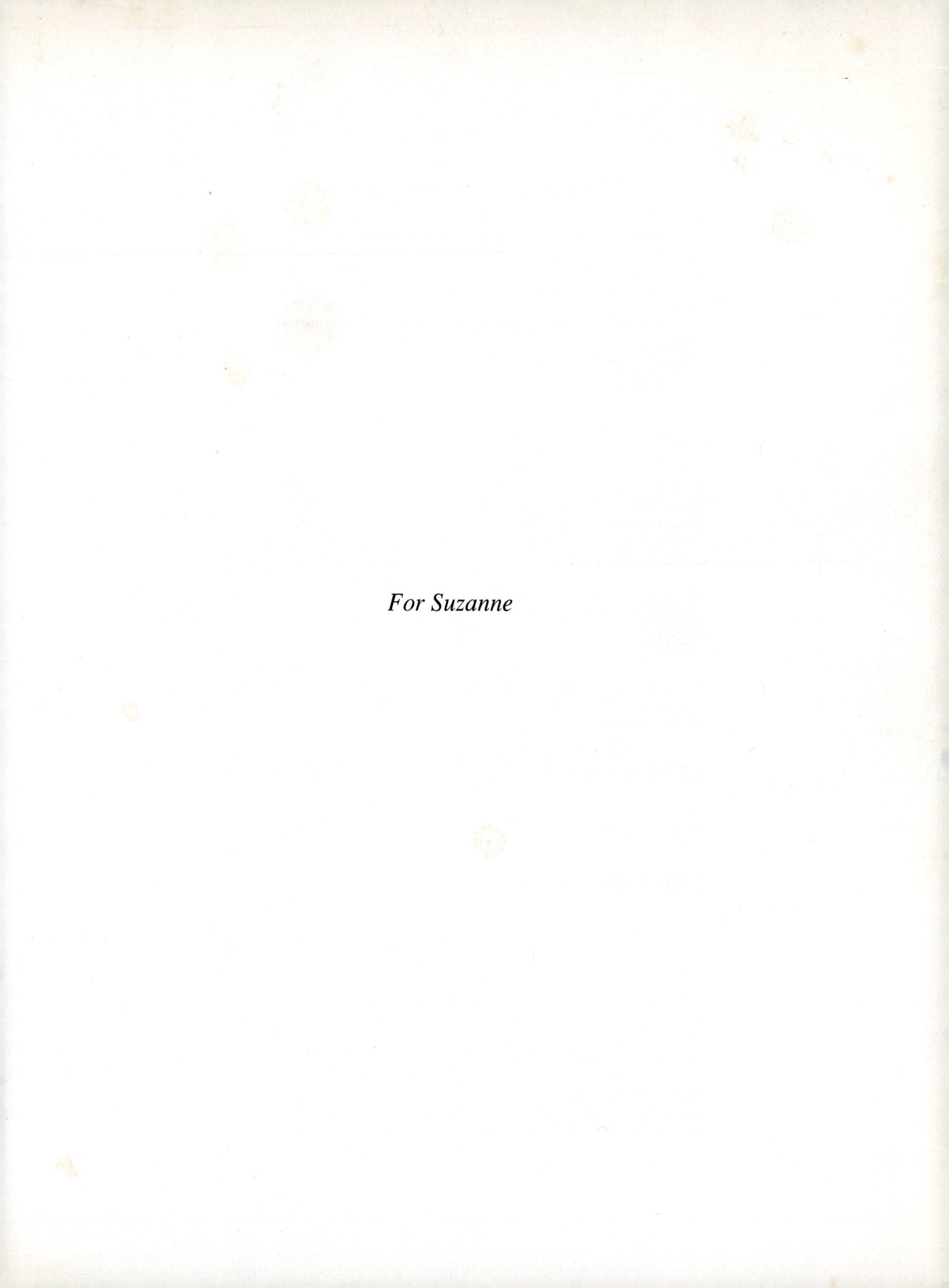

*For Suzanne*

# Contents

# Preface

*There is nothing permanent except change.*
Heraclitus, c. 500 B.C.

Hardly a day goes by without world news focusing on some event on the planet marked by change. These events include global warming, acid rain, ozone depletion, deforestation, and the elimination of species. These changes now affect more people than ever before in history. There is more concern by more people about the changes that are taking place on our planet than at any time since the first Earth Day in the early 1970s. Because of the widespread, sometimes severe impact that these changes have on the global community, there is ever-greater need to understand and forecast these changes.

This book is about global change and the nature of the impact. It is written as an introduction to global environmental change for the majority of college students that are not science majors. It incorporates basic science but is not based on an expansive knowledge of science and is designed to bridge the physical and social sciences. This book is designed for use in college- and university-level courses, where the core is an interdisciplinary approach to global environmental problems. These courses are now appearing in many different departments, including Earth science, environmental studies, geography, geology, and ecology.

Because many users of this book will not have had rigorous mathematical or scientific training, the book keeps mathematics, statistics, and physical principles at a basic level. Technical terminology is kept to a minimum as well. In all sciences, however, there is an essential vocabulary. To assist the reader with this terminology, I have included a glossary of key terms at the end of the book. To further aid the reader, chapter outlines and summaries are included. There are also sources for additional reading provided at the end of each chapter.

There is now worldwide concern over global environmental change. In the past several years delegates from many countries and from many academic disciplines participated in major international environmental conferences. The largest and most heralded of these conferences was the United Nations Conference on Environment and Development held in Brazil in 1992. Such conferences stress a need for inquiry into the natural processes of change and consider how the human species affects global environmental change, and vice versa. In the summer of 1994 an international conference on population was held in Cairo, Egypt. Delegates participated from most nations and from many private organizations.

Beginning in 1990 the U.S. government increased funding for research on global change. There are three major scientific goals in the global change program: (1) to establish long-range programs to watch the environment so as to detect changes with certainty and early in their development, (2) to increase efforts to understand the physical and biological processes producing the changes, and (3) to develop better models with which to predict future change. This book stresses the need for us to attain these goals.

In this book three aspects of global environmental change are addressed. The first is the nature of the basic processes that produce natural planetary change, including how Earth processes cause Earth's climate to change through time. Also dealt with are the geological processes that cause plate tectonics and the manifestations of earthquakes and volcanic eruptions. The second aspect is the effects of these changes on the human species. Any change in our environment has an impact on the human species. Changes in climate bring floods, drought, and unusually hot and cold periods, all of which affect agriculture and the world food supply. The third aspect deals with the ways in which the human species is altering the planet. The nature of global environmental change and the contribution of the human species to this change are both examined. There is no doubt that the human species changes the global environment. Often, it is difficult to sort out how much of the change is caused by human activity and how much is natural change.

What the future environment will be cannot be determined. That it will be different from what it is now is certain. It is important that all of us be aware of that change is occuring. This book is intended to illustrate the kind of changes which occur, what the impact of these changes will be, and how human action can alter the direction of some of these changes.

My thanks go to the reviewers of the manuscript that made so many positive suggestions. They are Anthony J. Brazel, Arizona State University; David W. May, University of Northern Iowa; John E. Oliver, Indiana State University; L. Michael Trapasso, Western Kentucky University; and Wayne M. Wendland, of the Illinois Water Survey.

I would also like to extend special thanks to D. Gordon Bennett and John E. Oliver for contributing chapters to the book, and to Jeffrey Patton for designing many of the maps.

Thanks are due to Paul Corey for the encouragement to produce the book, and to Robert McConnin, Executive Editor, for accepting the responsibility for developing the book, and to Rose Kernan for her very professional service as production editor.

chapter 1

# The Nature of Global Environmental Change

***CHAPTER SUMMARY***

Persistent Change
Rhythmic Change
Cyclical Oscillations
Short-Lived Events
Anthropogenic Change
Biological Response to Environmental Change

Change through time is a basic attribute of the planet. Earth has been undergoing constant change since it formed from a cloud of cosmic debris some 4.6 billion years ago. The changes that have been taking place, and are taking place, vary in form, size, duration, and areal extent. Days used to be shorter than they are now; the planet has been both warmer and colder than it is now; the magnetic poles of Earth have changed end for end. Mountain ranges have grown, then eroded away; ancient seas no longer exist, and biological species have appeared and disappeared. Even the sun, which supports life on the planet, is not a constant energy source.

Most changes that occur in our environment are not just random events. The changes are all the result of basic physical and biological processes operating on the planet. There are certain patterns that emerge in their occurrence through time.

## *PERSISTENT CHANGE*

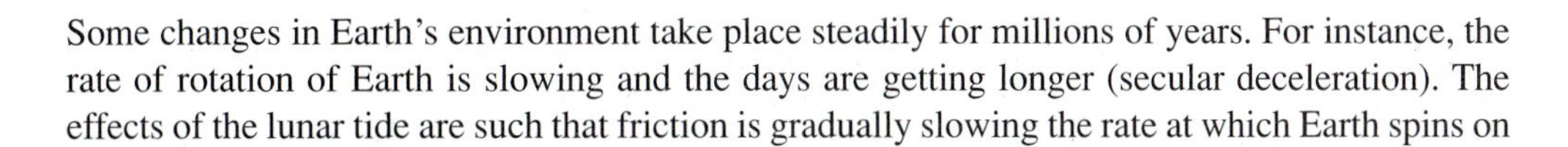

Some changes in Earth's environment take place steadily for millions of years. For instance, the rate of rotation of Earth is slowing and the days are getting longer (secular deceleration). The effects of the lunar tide are such that friction is gradually slowing the rate at which Earth spins on its axis. As the rate of rotation slows, the length of our day increases (Table 1.1). Five hundred seventy million years ago, there were 526 days in each year, and each day was only $16\frac{2}{3}$ hours long. Within the next 10 million years, a day will be lost out of our year. As Earth's rate of rotation slows, the moon moves farther away. As the rate of rotation slows, the length of our lunar month increases. Eventually, there will come a time when the day and month will be the same length, about 50 of our 24-hour days, and the effects of the tides will be less. The installation of power stations using tidal energy affects the rate of slowing in a minuscule fashion. The frictional effects of the dams and turbines make it harder for the water to flow in and out of the estuary and increase the drag on the tides.

The gradually slowing rate of rotation of Earth is an extreme example of a one-directional and apparently irreversible type of change. It occurred in the past, is now occurring, and will probably continue to occur in the future (Figure 1.1). Other examples that represent very slow rates of change are the changing shape of Earth and relative positions of the continents.

Earth is becoming gradually more spherical as its rate of rotation slows and the continents continually shift their positions relative to one another. They also change location relative to Earth's equator and poles.

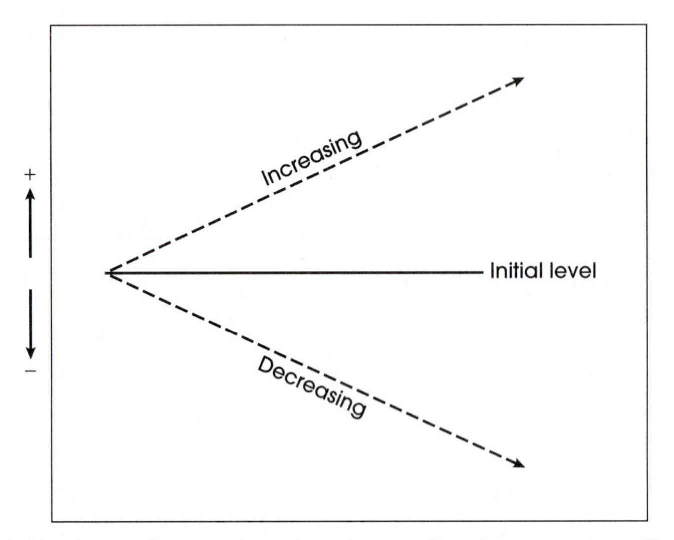


**Figure 1.1** Persistent change takes place in one direction over time. The variable may increase in quantity or decrease in quantity.

**TABLE 1.1** Changing Number of Days in the Year and Changing Length of the Day through Phanerozoic Time

| *Millions of years ago* | *Days in the year* | *Length of each day (hours)* |
|---|---|---|
| 65 | 371 | 23.6 |
| 245 | 389 | 22.5 |
| 360 | 421 | 20.8 |
| 570 | 526 | 16.7 |

## *RHYTHMIC CHANGE*

Some changes in the environment are rhythmic, or regular oscillations. In fact, periodic fluctuations, those occurring at regular, predictable intervals, include many of the well-known changes in the environment (Figure 1.2). There are several aspects of Earth's movement through space that result in many of the regular undulations that appear in the world around us. For example, the rotation of Earth on its axis produces day and night and the primary tides. The revolution of the moon about Earth produces the 28-day lunar month. The revolution of Earth about the sun produces the seasons. These events, so familiar even to children of a very young age, cause a virtually limitless number of associated regular oscillations in the immediate Earth environment. The majority of these oscillations are unnoticed by the average person. The following paragraphs provide some examples of these periodic oscillations.

The moon forces rhythmic changes in many aspects of Earth. The size and position of the moon are such that the gravitational force on the side of Earth closest to the moon is 7 percent greater than on the point opposite. Earth's crust, atmosphere, and seas are influenced by this gravitational force. The moon exerts a pulling force on the elements on the side of Earth toward the moon. The seas raise up by 1 meter or more. As Earth rotates on its axis, constantly presenting a different face to the moon, the upward surge of water moves around the planet. Most locations experience two high tides and two low tides a day. One high tide occurs about the time the moon is directly overhead, and the other high tide occurs when the moon is 180° away, on the opposite side of Earth.

The sun also exerts gravitational pull on the planet. The moon is much closer to Earth than the sun and exerts about twice the gravitational force. The attraction of the sun thus alters the tides associated with the moon but does not create its own.

Other regular oscillations also influence our lives (Table 1.2). Lunar eclipses are an example, as is the appearance of a particular species of cicada, mistakenly called a 17-year locust. The reappearance of Halley's comet every 76 years is a regular event; it last approached the sun in 1986.

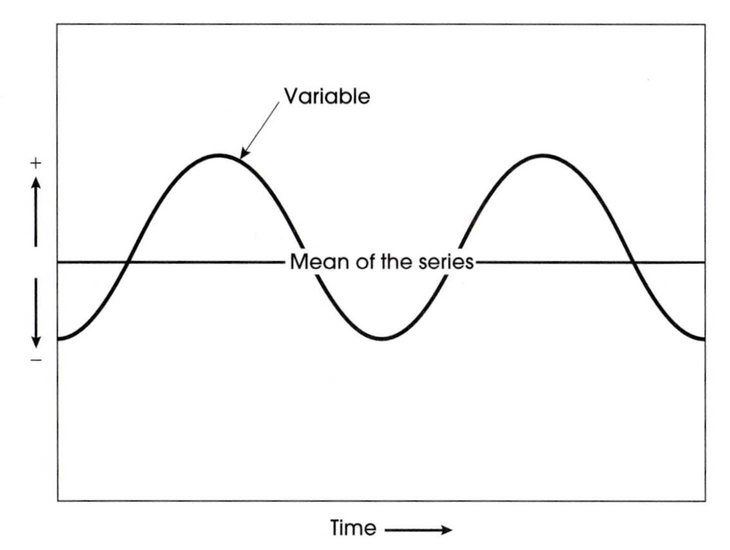


**Figure 1.2** A periodic phenomenon repeats itself at regular intervals and at regular amplitudes.

**TABLE 1.2** Periodic Changes in the Global Environment

| | |
|---|---|
| Primary tides | 12 hours 24 minutes |
| Day | 24 hours |
| Spring and neap tides | 14 days |
| Lunar tides | 28 days |
| Year | 365 days |
| Axial precession | 23,000 years |
| Axial tilt | 41,100 years |
| Earth's eccentricity | 100,000 years |

A much longer rhythmic movement is the precessional movement of Earth's axis. Earth spins like a top and the orientation of the axis relative to our Milky Way galaxy changes through time. It takes about 26,000 years for the axis to make one circle. It is for this reason that the position of the star Polaris changes with time. The star is never exactly in line with Earth's axis, and the distance it deviates changes as the axis makes its slow turn in space. Polaris will be most in line with Earth's axis in A.D. 2100, when it will be off only half a degree. It is now nearly a degree away from alignment with the axis.

## CYCLICAL OSCILLATIONS

Many environmental phenomena repeat themselves at irregular intervals and with varying intensity; that is, they are not strictly periodic. These phenomena, such as droughts, recur, but it is not possible to predict when they will occur. A cyclical phenomenon is one that repeats itself, perhaps at some average interval of time, but the actual interval between episodes varies from the average. For example, a phenomenon may have an average interval of seven years, but successive intervals, including four episodes, may actually be eight years, four years, six years, and ten years. Another variable associated with cyclic phenomena is the fact that not only do events vary with time, but the size of events varies (Figure 1.3).

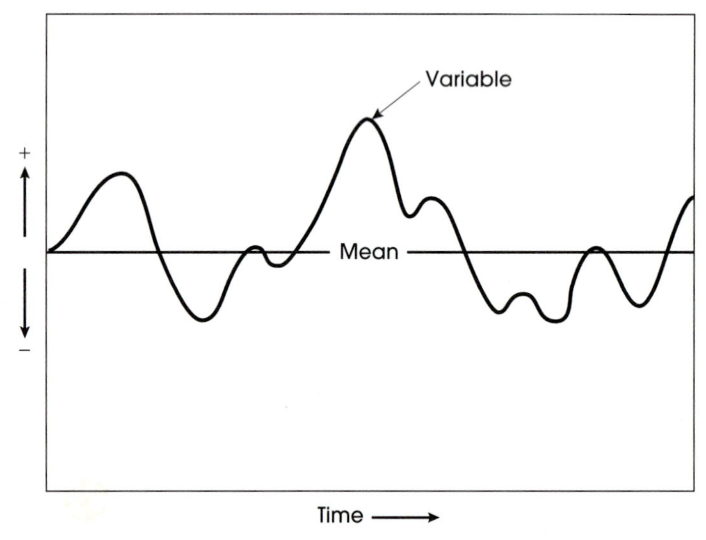


**Figure 1.3** An aperiodic or cyclical variable repeats itself, but at irregular intervals, irregular amplitude, or both.

Irregular oscillations are common in Earth's environment. References to such changes appear in the earliest written records of the human species. The seven lean years followed by the seven fat years alluded to in the Bible is a reference to a moisture cycle. Rainfall data for parts of Earth show precipitation cycles with a length averaging 11 years. Some scientists say that these cycles are caused by sunspots.

Geological records, biological records, and historical records all show that our atmosphere is never the same for very long periods. On the basis of this information it is recognized that there is really no such thing as normal weather or climate. Weather oscillates in an irregular manner between warm and cold years and wet and dry years. Years of warm or cold weather often cluster together, as do wet years and dry years. Implicit in this clustering is the knowledge that weather is not a random process.

The time scale for changes in weather events varies, and this is the basis for distinguishing between weather and climate. Weather is the condition of the atmosphere at any given point in time and space. Not only does the weather at any given place change with time, but different places experience different weather. Snow is part of the weather in some areas but not in others. Tropical cyclones are a type of weather experienced in parts of the tropics and subtropics but not in all parts of the tropics and not in polar regions. Weather changes constantly with time, but the intervals between changes are short, measured in hours, days, or weeks.

Climate consists of the regime of weather at a given place over a longer period. It includes the most frequently occurring types of weather as well as the infrequent types. While weather changes rapidly with time, climate is more stable and changes occur over longer periods. It may take decades or centuries to detect true changes in climate. One of the important scientific questions today is whether or not the earth's climate is changing in response to human-induced changes in atmospheric chemistry. At this time there is no clear answer to this question.

There are short-term oscillations in the environment. These are short swings toward extremes in which the time element is 1 to 10 years. These short oscillations are verified by actual measurements. There are many variations in weather over periods of months or several years. A cluster of unusually cold winters occurred in North America in the years 1977–1979. The events popularly refered to as El Niño are another example.

Populations of some biological species oscillate in an irregular manner. One species that exhibits such behavior is the lemming, a common rodent of the arctic tundra. The species tends to oscillate in a cycle that averages three to four years in length. One year the population will be extremely high and the following year the population will crash. There is then a slow buildup of the population for one to three years and then another crash. The cycle may be related to the food supply. The lemming population grows until it consumes nearly all the edible food.

The following summer, when a still larger population appears on the landscape, there is simply not enough food to go around. Large numbers of lemmings die of starvation. It is the overrunning of the food supply that causes the massive migrations associated with lemmings. What causes this cycle is not known for certain, and the cycle does not appear in other arctic species.

When the population of any species varies too widely, it is possible that in an extreme downturn the population will become too small to perpetuate itself and the species will die out. Extinction occurs, of course, when the population size reaches zero. However, extinction can occur in any species when the population size drops below some minimum threshold number needed to reproduce enough offspring to sustain the species. In some species this threshold number can be very large. Thus it is possible in one wild fluctuation in a species population that extinction becomes certain.

Intermediate-length oscillations range from 10 to 1000 years in length. Such oscillations are long enough to put many of them beyond proof by direct measurement. In North America, records of climatic data begin around 1800. This is the beginning of historic time in the direct measurement of climate. It is only enough time to examine short-term oscillations and a few of intermediate length. Simultaneous recording of weather data on an international scale has taken place only during the past 100 years. Systematic observations of the upper atmosphere and of water temperatures at the surface of the ocean have existed only since the 1950s.

Long-term oscillations are those where the extremes occur at intervals of more than 1000 years. Records show there have been many irregular variations of large amplitude in the past. In fact, the data suggest that the longer the available record, the more extreme the climatic variations. The short record of instrumental measurement simply does not show the large climatic variations that occur. There is a fundamental error in using 30-year climatic normals, or in considering instrumental records of 100 years in length as being representative of climate. Some environmental fluctuations are thousands or tens of thousands of years in length. An example of the longer oscillations are those associated with the ice ages.

## *SHORT-LIVED EVENTS*

Some events occur in our environment that last a span of days or even seconds. These events often are sporadic in space and time. They represent major deviations from normal conditions and are frequently violent events that result in disaster for the human species (Figure 1.4). These events may be atmospheric, hydrologic, geologic, biologic, or come from space. They are also extremely difficult to predict. The duration of short-lived events varies over a wide range. Lightning strokes last milliseconds and are among the most instantaneous of phenomena. Floods may last for a few hours or for weeks. Table 1.3 provides the time span within which each phenomenon occurs. In each case the phenomenon is listed under the shortest time span in which it occurs. Some persist for longer periods and thus spread across more than one class.

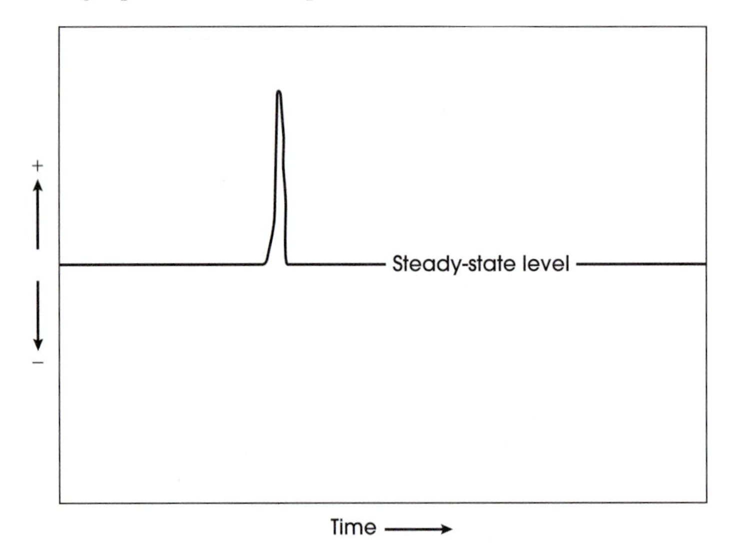


**Figure 1.4** A short-lived phenomenon is a sudden departure from the normal or mean. The system usually returns to its original state but in some cases goes to a new mean state.

**TABLE 1.3** Time Scale of Short-Lived Phenomena

| *Seconds* | *Minutes* | *Hours* | *Days* |
|---|---|---|---|
| Lightning | Tornadoes | Blizzards | Tropical cyclones |
| Earthquakes | Hailstorms | Riverine floods | Cold waves |
| Landslides | Tsunamis | Volcanic eruptions | Insect population explosions |
| Avalanches | | | |
| Meteorite falls | | | Heat waves |
| Atmospheric fireballs | | | |

While short-lived phenomena are similar in some respects, they differ widely in others. The size, frequency, and spatial and temporal characteristics are different. Some short-lived phenomena have a similar size with each occurrence. Tropical cyclones, although varying in size and intensity, have wind velocities and central pressures that are within a range of a factor of 1. Other events vary widely in size. Earthquakes are an example of this class. The range of size that human beings can feel varies by a factor of 1 million. There is the same kind of variation in flood magnitude. The longer the period of record, the greater the chance of having a more extreme case of a short-lived phenomenon.

The frequency is a measure of how often an event of a given size may occur, or it may be how often the event occurs if there is little difference in size from one event to the next. The concept of frequency is far more important in the analysis of some events than others. Data show that the frequency of short-lived phenomena such as floods and earthquakes is increasing. It is virtually impossible to determine if this is in fact the case since records of these events exist for only a limited time. Records of floods on the Nile River go back many centuries, but this is an exception record. Few nations began to compile data on extreme events until recent years. It is only the giants of extreme events for which there are records that have existed for more than a few decades.

Human interference in the environment may have increased the frequency of some phenomena. Impounding water in large reservoirs triggers earthquakes, as does injecting fluids into the ground. In still other cases, withdrawal of oil and water from the ground causes earthquakes. Many flash floods are caused by the collapse of dams and reservoirs. The frequency of hailstorms and landslides in some areas is altered by human activity. Human activity adds to the difficulty of determining what the natural rates of occurrence of some phenomena really are.

Geographical distribution refers to the pattern of a phenomenon over the area in which it occurs. The areal extent ranges widely, from a lightning bolt on one hand, to a tropical cyclone on the other. Some phenomena, such as floods and avalanches, have very well-defined limits. Others, such as hurricanes, are less well defined on the margins.

The temporal pattern of short-lived phenomena also varies widely. Although most have a random element, there may be daily, seasonal, or annual fluctuations in the probability of an event. There may be cyclical patterns to events of undetermined characteristics.

## ANTHROPOGENIC CHANGE

Human activity on Earth has reached such an intensity and such a large spatial extent that it is a major agent for changing the planet. Much of the human impact on the environment is a direct consequence of the number of people on the planet. A second element in the impact is the unequal distribution of the species. Within the past 20,000 years, and perhaps as far back as 100,000 years, the human species has grown sufficiently in numbers and technology to produce global change. Within the past several hundred years the rate of impact has grown at an increasing rate. This is a result of the rapid growth of technology and the power this technology has provided the species. The past 50 years has seen the world population double and world economic output quadruple. In the same 50 years human beings traveled to the moon and back, and human machines traveled across the solar system and into outer space.

Every living organism changes the environment in which it lives. Organisms remove from the environment what they need for their existence. In the process they change the habitat in which they live and perhaps destroy other organisms. The first green plants to produce oxygen started a process that led to an atmosphere rich in oxygen. Many species that had survived in a carbon dioxide–rich environment perished in the oxygenated world.

The human species changes its habitat, and in fact has changed the planet to an extent that no other species before it has done. From the modest beginnings of the species the numbers have grown at an ever-increasing rate. The primary means by which the species has altered the environment has been through the search for food for the rapidly growing population. The search for food and other goods for the species has grown at an even faster rate than the population.

The explosion of the human population on the planet has brought with it many other changes resulting from the needs and desires of the species. Much of the human impact on the environment is due directly to the number and distribution of humans on the planet. As humans have changed from being hunters and gatherers of wild foods to being exploiters of a wide variety of earth materials, they have altered the earth system both directly and indirectly. In some cases human activity takes the form of direct intervention in natural process. Some examples are the injecton of greenhouse gases into the atmosphere, deforestation, soil erosion, and the elimination of species. Humans also intervene in natural systems indirectly through regulatory systems, economic and trade relationships, taxes, and subsidies. Local, state, national, and international policies also affect environmental change indirectly.

There is no longer any major area on or near the surface of the earth in which the species has not left its mark. Humans have been to the tops of the highest mountains. The deepest trenches in the world ocean contain debris that has been dumped intentionally or accidentally. The Arctic and Antarctic Oceans carry persistent pesticides in measurable quantities. The atmosphere to its highest levels carries radioactive debris from nuclear weapons, and now the moon and some of the planets carry wreckage of space vehicles from Earth. The presence of human beings is ubiquitous. The extent to which we have altered the environment varies over the Earth, but the presence of human modification is everywhere.

The extent and variety of human-induced changes has grown at a geometric rate and at a rate greater than that of the growth of the population. The human population has grown to such an extent and become so concentrated in some areas that the production of waste materials has surpassed the capability of the local environment to absorb or destroy them.

Technology has increased the amount and variety of wastes dumped into the environment. Hazardous wastes include synthetic products such as plastics, chlorinated hydrocarbons (DDT, dieldrin, and related compounds), and chlorofluorocarbons. The natural environment has no means, at least no rapid means, of disposing of these synthetic wastes. The creation of new chemical products and the introduction of huge quantities of chemicals into the Earth system has altered the chemical balance of the entire Earth surface. Earth's atmosphere, ocean, and even the crust have been altered chemically. Since water and the atmosphere are highly mobile and circulate over the planet, changes wrought in them spread over the entire Earth.

The biosphere is also rapidly changing. Natural vegetation is being altered at a very high rate. Crises have developed for many species of flora and fauna existing during the period of human habitation of Earth. These crises have led to extinction for many of these species. Overspecialization or evolutionary problems have caused some to disappear. For many, human beings produced the crisis and forced extinction. As *Homo sapiens* continue to clear land for agriculture and continue to use more chlorinated hydrocarbons, the potential for extinction of other species rises at a startling rate. Land-use changes resulting from human activity, particularly habitat fragmentation and deforestation, are part of one form of irreversible global change. This is the loss of biodiversity through the extinction of plant and animal species. The current rate of loss of species exceeds that of any past extinction episode.

The persistent changes that are taking place now are in some cases irreversible and in some others reversible. Extinct species are permanently extinct. Although it is possible to replant large areas of Earth, natural rates of soil development cannot replace lost soil in a human lifetime or even several lifetimes. All these changes have direct and indirect socioeconomic impact at some regional level.

Today, human activity changes natural environmental processes through the long-term and collective effects of individual actions. The scope of these actions depends on the social and economic systems that the human species has developed through time. In the space of just a few centuries, an instant of time in geological terms, the species has altered Earth's surface and atmosphere and the world ocean. The changes that have taken place in this short period are phenomenal.

Analysis of global change must now include the explosive growth of human numbers and the even larger exponential growth of human activity. It must include the social systems that the human species has developed and the variation in these social systems from place to place.

## *BIOLOGICAL RESPONSE TO ENVIRONMENTAL CHANGE*

Life-forms have responded to environmental change through the process of evolution. The periodic changes from day to night and season to season have aided in developing species adaptable to a changing environment. Species in high latitudes have had to adapt to wide fluctuations in temperature, day length, and food supply. In the wet-and-dry tropics the alternation of seasons forces a different kind of adaptation. When some unusual change occurs in the environment, it may produce an altered environment with a different set of climatic or other conditions. After each major natural environmental change, the life-forms that had adapted to earlier conditions either adapted or disappeared. It is possible, and highly probable, that many species will not survive the present changes taking place on the planet.

The impact of global environmental changes on human beings can take many forms. It can be measured in human stress. Both physical and mental health are impaired by stress associated with environmental changes. Events such as lightning, earthquakes, volcanic eruptions, landslides, and floods sometimes take lives instantly. Lives are lost by disease, injury, or starvation. Changes in climate can mean the loss of a means of livelihood, especially in agricultural economies. Loss of crops or animal products may result in malnutrition, disease, or starvation.

There may be long-lasting and far-reaching effects of environmental change beyond fatalities alone. The extent of the impact depends on a variety of factors, including (1) the size of the event, (2) the number of people affected, and (3) the cultural stage of the people affected.

Earth is always changing due to the natural physical and biological processes that operate in the universe. Over 4.6 billion years these processes have molded the planet, and today the same natural processes operate. However, to understand the planet as we live on it today, we must include changes brought about by the human species. Those changes brought about over the past several centuries, and those forecast for the next decades and the twenty-first century and beyond, are large. In the future, the historic record will show the current centuries as a period of catastrophic change on the planet—just how catastropic only time will tell.

To place environmental change in proper perspective, we must place in proper perspective the time scale in which changes occur. People often judge time by their own experience. For example, atmospheric temperatures vary through the day, sometimes by many degrees, and we are aware of these changes as they affect our physical comfort. However, over a period of weeks the changes brought by different weather systems become more important than the diurnal temperature changes. Over a period of a year or more the daily temperatures and cyclical changes of rainy and dry spells become less important than the seasonal changes. In the span of a lifetime, only the years or clusters of years that represent drought, or unusual cold or heat, are memorable. So it is with most environmental change. An event that looms large at the moment often turns out to be an average event in the longer scheme of things.

## *SUMMARY*

The environment of Earth has always been undergoing change. There are different patterns and frequencies of change through time that interact to keep the global system ever changing. There are persistent changes such as the slowing of Earth's rate of rotation and movement of the crustal plates. There are rhythmic changes, such as the seasons and tides. Other changes recur through time but with an irregular size and at irregular intervals. Short-lived events often produce natural disasters. Now global change is being brought about by the human species. All of these types of change to the planet occur simultaneously. The result is a complex planetary system in which change is not easy to forecast. As the human population grows, the impact of unexpected changes increases in magnitude and geographic extent.

## BIBLIOGRAPHY

Bryant, E. A. 1991. *natural Hazards*. Cambridge, England: Cambridge University Press.

Burton, I., R. W. Kates and G. F. White. 1993. *The Environment as Hazard*. Second Edition. New York: Guilford Press

Committee on Earth and Environmental Sciences. 1990. *Our Changing Planet: The FY 1991 Research Plan* of the U.S. Global Change Research Program.

Corell, R. W. and P. A. Anderson. 1991. *Global Environmental Change*. Berlin: Springer-Verlag.

Mannion, A. M. 1991. *Global Environmental Change*. Harlow: Longman Scientific and Technical.

Smil, V. 1993. *Global Economy: Environmental Change and Social Flexibility*. New York, NY: Routledge, Chapman, and Hall

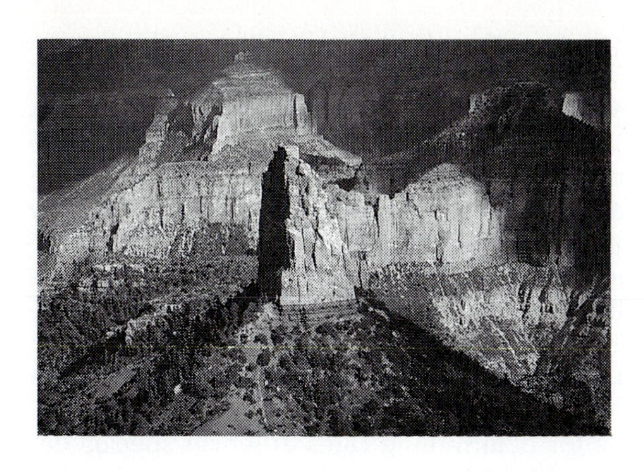

# part I

# GEOLOGICAL RECORD OF GLOBAL CHANGE

Precise measurement of global environmental change is available for only a very small bit of Earth's history. The most and best information about the planet is available for the immediate past. The further back in time one looks, the less information there is and the less precise it is.

The present global system represents the latest result of processes that have operated throughout Earth history. The geological period covers the entire span of Earth history from the formation of the planet to the present. The geological record contains data on global processes and the effects of these processes in the past. It adds a time element to global changes not available in historic time. It also yields information on processes that operate under climatic, geographic, and tectonic settings unlike the present or recent past.

Earth history broadens the perspective on the natural limits and scales of global change. Geological data from the early part of Earth's history provide information about changes that took place over a very long period. In some cases the time involved is millions of years. Continents moved about on the planet and continental glaciers formed and melted away.

Evidence from the past shows that there have been different steady-state conditions of Earth's circulation systems of the atmosphere and hydrosphere. Earth's climate may have, and probably did, shift rapidly between warm and cool conditions. It appears further that at least some of these fluctuations took place without any external influence. They were the result of changes within our planetary system. There are also indications of changes in climate on time scales of 10,000 to

100,000 years that are associated with variations in the orbital geometry of the planet.

The geological record contains critical information about environmental extremes. Geologic data show changes that were of tremendous size. We now know the general nature and time of some of these extreme and abrupt changes. Large objects from space struck Earth on many occasions. Earth's magnetic field reversed itself a number of times. Evidence remains of only the major events. Information no longer exists about multitudes of changes. Many must have been extremely significant at the time, but later events destroyed the evidence of these changes.

Geologic time provides a basis for determining rates at which species develop and existing species become extinct. Earth history also provides a record of times when life on the planet was greatly stressed. Rapid changes in the past resulted in major migrations of living organisms and repeated mass extinctions. Some of these mass extinctions eliminated the majority of species living on the planet.

The natural variability of the global environment also provides a background against which human impact can be assessed. One of the difficult questions for science today is to determine what form and how much of global environmental change today is attributable to the human species.

chapter 2

# Early Geologic Time: The Precambrian Era

***CHAPTER SUMMARY***



A perspective on the type, amount, and time of change is very important in a book about global environmental change. We must recognize that the time in which we live is not typical of conditions over most of Earth's history. Neither the climatic environment nor present living organisms are typical of those of most of Earth's existence, during which the planet was much warmer than it is at present. Earth's atmosphere has trapped enough Earth radiation to keep the planet warmer than it is today. This predominately warm environment was punctuated with several episodes when the planet cooled well below the normal warm conditions. During these times the planet cooled enough that continental ice sheets formed on the continents.

In its early state there was no solid crust. The crust that now makes up the continents did not form until almost 1 billion years after the planet first formed. Since the time the continental crust formed, it has been undergoing continual change. From the time of their formation, continental plates have undergone extensive periods of mountain building and erosion. The continental plates have also merged into a single landmass and then split several times.

Earth is unique among the planets. It is the only planet that has abundant water in all three states: solid, liquid, and gas. It is the only planet with a warm, moist climate ideal for living organisms as we know them. In the beginning there was no life, but living organisms emerged early in geologic time. Once living organisms appeared, their numbers and complexity grew rapidly. It was not a steady uninterrupted progression, however. There have been extreme events on the planet that resulted in massive extinctions of living organisms. Planetary changes of monumental proportions caused these mass extinctions. One such episode eliminated at least 90 percent of all living species on the planet.

Geologists have divided the history of Earth into eons, eras, and many smaller periods and epochs. Table 2.1 presents the structure of the time units and samples from the geologic timetable. Table 2.2 lists these time units and the approximate length of each. Eras, periods, and epochs are differentiated on the basis of life-forms. Each boundary is a point in time when there was a drastic change in the global environment. Each boundary coincides with an important step in the evolutionary process.

It has been common practice to consider geologic time as comprising four eras, with the addition of eons in more recent years. Table 2.3 shows the relationship between eons and eras, and

**TABLE 2.1** Structure of Geologic Time Units

| *Unit* | *Example* |
|---|---|
| Eon | Phanerozoic |
| Era | Cenozoic |
| Period | Quaternary |
| Epoch | Holocene |

**TABLE 2.2** Geologic Timetable

| *Era* | *Period* | *Epoch* | *Beginning (millions of years before the present)* |
|---|---|---|---|
| Cenozoic | Quaternary | Holocene | Present |
| | | Pleistocene | 1.6 |
| | Tertiary | Pliocene | 5.3 |
| | | Miocene | 23.7 |
| | | Oligocene | 36.6 |
| | | Eocene | 57.8 |
| | | Paleocene | 66.4 |
| Mesozoic | Cretaceous | | 144 |
| | Jurassic | | 208 |
| | Triassic | | 245 |
| Paleozoic | Permian | | 286 |
| | Carboniferous | | 360 |
| | Devonian | | 408 |
| | Silurian | | 436 |
| | Ordovician | | 505 |
| | Cambrian | | 570 |
| Precambrian | | | 4600 |

**TABLE 2.3** Eons and Eras of Geologic Time

| *Eon* | *Era* |
|---|---|
| Hadrian | |
| Archean | Precambrian |
| Proterozoic | |
| | Paleozoic |
| | Mesozoic |
| Phanerozoic | Cenozoic |

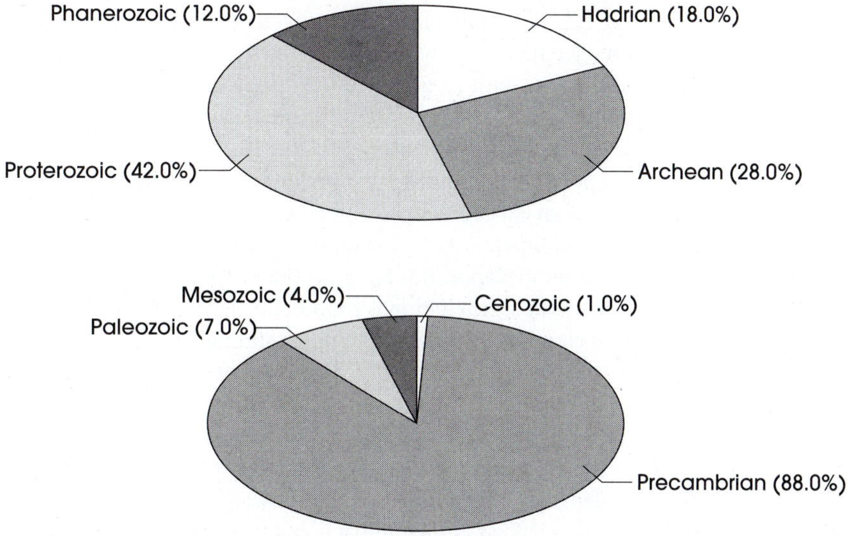

**Figure 2.1** Percentage of geologic time in each of the four eras. Each successive era is shorter than the previous one. Most of geologic time is in the Precambrian.

Figure 2.1 illustrates the relative length of the four eons and four eras. The earliest and longest geological era is the Precambrian, which spans 88 percent of Earth's history and includes three eons. It includes the long ages of time about which there is the least known about conditions on the planet. Evidence of what took place on the planet in the Precambrian is skimpy. Although there is less information about this part of Earth history, we now know some important events that took place. Much of what we now accept as fact is the result of intelligent guesswork.

## *ORIGIN OF PLANET EARTH*

There is no single theory accepted by everyone that explains the origin of Earth. Among the various theories, one is the most likely alternative. This theory suggests that the sun and the Earth, together with the other planets and their moons, formed when a portion of a cloud of interstellar gas and dust began to coalesce. Over time, particles of gas, dust, and ice began to collect into planetesimals, or little planets. As these formed and grew larger, the gravitational forces associated with them increased. More and more of the mass of particles gathered until only the largest masses remained. The process cleared a large space of debris. From this process our solar system emerged. The age of the oldest rocks found on Earth, the age of meteorites, and the age of rock samples from the moon are similar. They confirm a common origin about 4.6 billion years ago. All evidence suggests that Earth began as a molten mass. The temperature at the surface was more than 8000°C (14,400°F). Some of the gases present during this initial period, mainly hydrogen and helium, were very hot. Their molecular velocities were so high that some overcame the gravitational pull of Earth and escaped to space.

## *BOMBARDMENT FROM SPACE*

Planetesimals of all sizes bombarded Earth continually until about 3.8 billion years ago. These contained large amounts of ice and carbon compounds. It may have been that this debris from space provided the planet with the basic materials for our atmosphere and oceans. The pitted surface of the moon, with its many craters, is visible evidence of the bombardment that took place. The continual recycling of Earth's oceanic crust and erosion of the continents erased most of the evidence of this bombardment.

It is possible that during this period of consolidation of the planet, the moon formed from the impact of an extraterrestrial object with Earth. The theory postulates an object traveling at a speed of 11 kilometers per second, with the collision vaporizing part of the object. Heated to 6600°C, the jet of vapor, traveling at a high speed, escaped the planet. The cloud of vapor remaining in Earth's gravitational field condensed into a mass of small rock fragments orbiting Earth. This cloud of debris fused together through collisions to become the moon. The chemistry of the moon supports this theory. The composition resembles that of Earth. Missing from the moon are the most volatile chemical elements—these escaped into space.

Following the initial period of bombardment, planetesimals continued to reach the planet but at far less frequent intervals. The discovery of very large impact craters on Earth has provided direct evidence of this bombardment. They also show the size of some of the objects that struck Earth. The two largest known impact craters measure 140 kilometers in diameter. One is north of Lake Huron near Sudbury, Canada, the other near Johannesburg, South Africa. Both are about 1.9 billion years in age. They are so old that none of their crater rims remain. Only the disintegrated rock beneath the point of impact serves as evidence of the collision. There also exists the remains of a crater 13 kilometers wide located 200 kilometers south of Chicago which formed about 300 million years ago. Barringer Crater, or Meteor Crater, in Arizona is much smaller and much younger than those just noted. It contains remnants of the iron-nickel meteorite that formed the crater. The crater is only about 1.2 kilometers in diameter and 150 meters deep. Its estimated age is about 50,000 years. The energy released by that impact was only 1/10,000,000 that of the larger asteroids that produced the craters in Canada and South Africa.

About 120 craters are considered to be of meteoric origin. So far none has been shown to be older than 1.9 billion years. There is evidence, in the form of particles containing high levels of the element iridium, that meteoric impacts occurred before then. Iridium occurs in only small amounts in Earth's rocks but is more abundant in meteorites. There are beds of rock in South Africa that contain high concentrations of iridium. The age of these rocks is 3.25 to 3.45 billion years. The layers of iridium-enriched rock indicate four large meteorite strikes that spread fine particles around the Earth. The impacts could have occurred any place on Earth, as wind distributed the fine particles quite widely. South Africa is a good place to look for such deposits because rocks there are among the oldest on the planet.

An estimated 100 billion tiny meteorites strike our atmosphere each day, their average size being that of a grain of rice. This rain of minute particles is not unusual or extreme. Most particles burn from the heat generated by friction with the atmosphere. Most burn before the particles get within 50 kilometers of the surface. It is probable that all meteorites of less than 100 grams burn in the atmosphere. The event is noteworthy when a particularly large meteorite flashes through the air as a fireball or hits Earth. Since most of Earth's surface is water, most meteorites that reach the surface strike the sea. More than 130,000 meteorite fragments exist out of the multitudes that have hit Earth. Most that strike land are too small for identification by the layperson, and even some of the larger ones look like ordinary rocks.

Not all meteorites that come near Earth are small. On March 23, 1989 a meteorite 330 meters in diameter came within 90,000 kilometers of Earth, a relative distance twice that of the Earth to the moon. This asteroid was perhaps a tenth the size of the object that produced the Tunguska Event of 1908. If it had collided with Earth on a continent, there would have been major damage.

## PLATE TECTONICS

Modern geophysical research has provided a unifying theory for the old idea that the continents drift over Earth. The present positions of the world's landmasses are only temporary in the long-term evolution of the continents and oceans. The early bombardment of the planet and the decay of radioactive elements heated the planet to the melting point. The planet was molten for the first few hundred million years. While it was molten, mineral separation occurred, during which process heavy elements such as iron sank toward the center of the planet. Earth's iron core formed from this process (Figure 2.2). The lighter materials migrated to the surface. Eventually, Earth cooled enough for the lightweight material to solidify into a rocky crust. It is this lighter-weight mass that now forms the continents. It is unlikely that any new continental crust is forming at present.

The continental crust rests on a more dense mass called crustal plate. Crustal plates are more dense than the continental material. They are made up largely of basalt, which is less dense than the material beneath. Beneath the crustal plate lies the mantle, which is plastic and contains convection cells that move ever so slowly. The brittle continents lie on top of the mantle and move in response to movement in the mantle beneath. The crust more or less floats on top of the plastic mantle.

In response to motion in the mantle, the solidified crust cracks into pieces called plates. There are many such plates, or pieces of the crust, that make up the outer layer of Earth. They are in constant motion. Each continent lies on one of these crustal plates. Over time the plates break into smaller pieces or fuse into larger ones. There have been times when all the pieces of continental crust have migrated together to form a single large landmass. This process of motion, and the creation and destruction of crustal plates, is known as plate tectonics. Figure 2.3 sketches some of the principal plates. Plate tectonics is one of the major processes of long-term regional and global environmental change. It has been a factor in changing the climates of the continents, in opening and closing seas, and in building mountain ranges.

## TYPES OF PLATE BOUNDARIES

At the boundary between two plates one of several possible processes may take place (Figure 2.4). One is that the two plates may move laterally past each other. This is a transform fault boundary. When this happens, friction between the two plates causes the edges of the plates to be torn, resulting in many cracks or faults in the crust. This is what is taking place in California. Part of

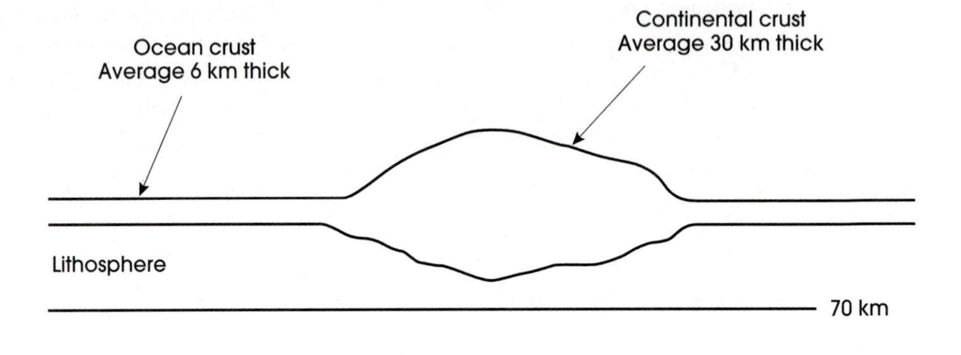


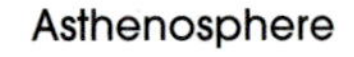


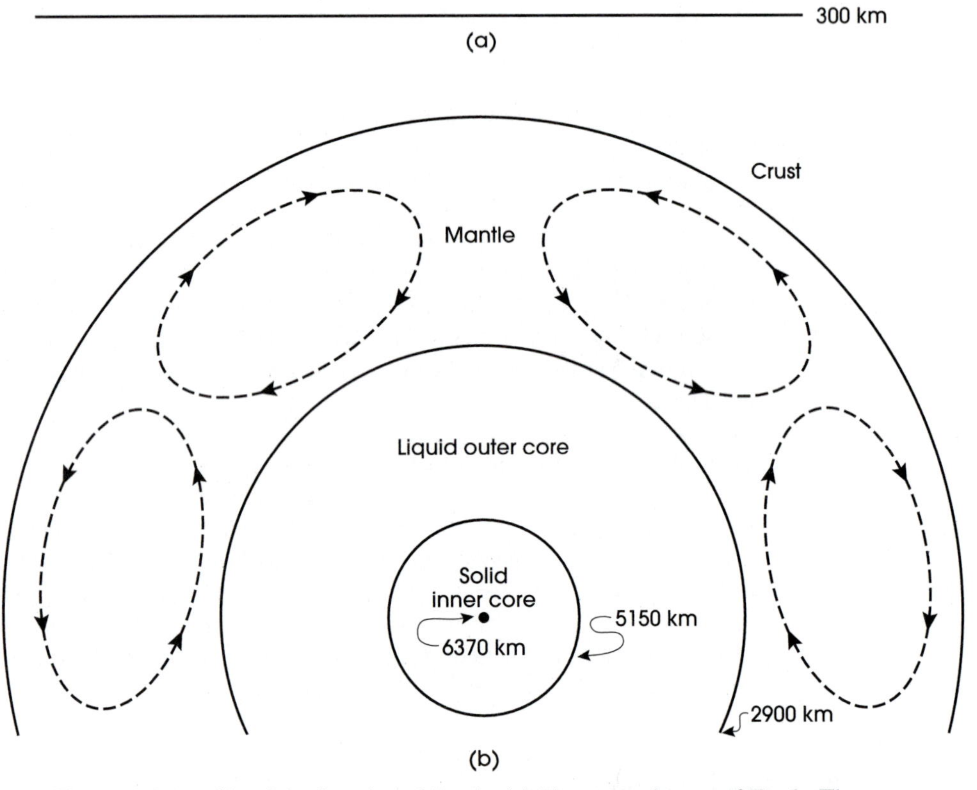


**Figure 2.2** Profile of the interior of Earth. (a) The upper layers of Earth. The crust and lithosphere are brittle and move on the plastic asthenosphere. (b) Profile through Earth. The motion of the crustal plates is due to the slow convection in the mantle.

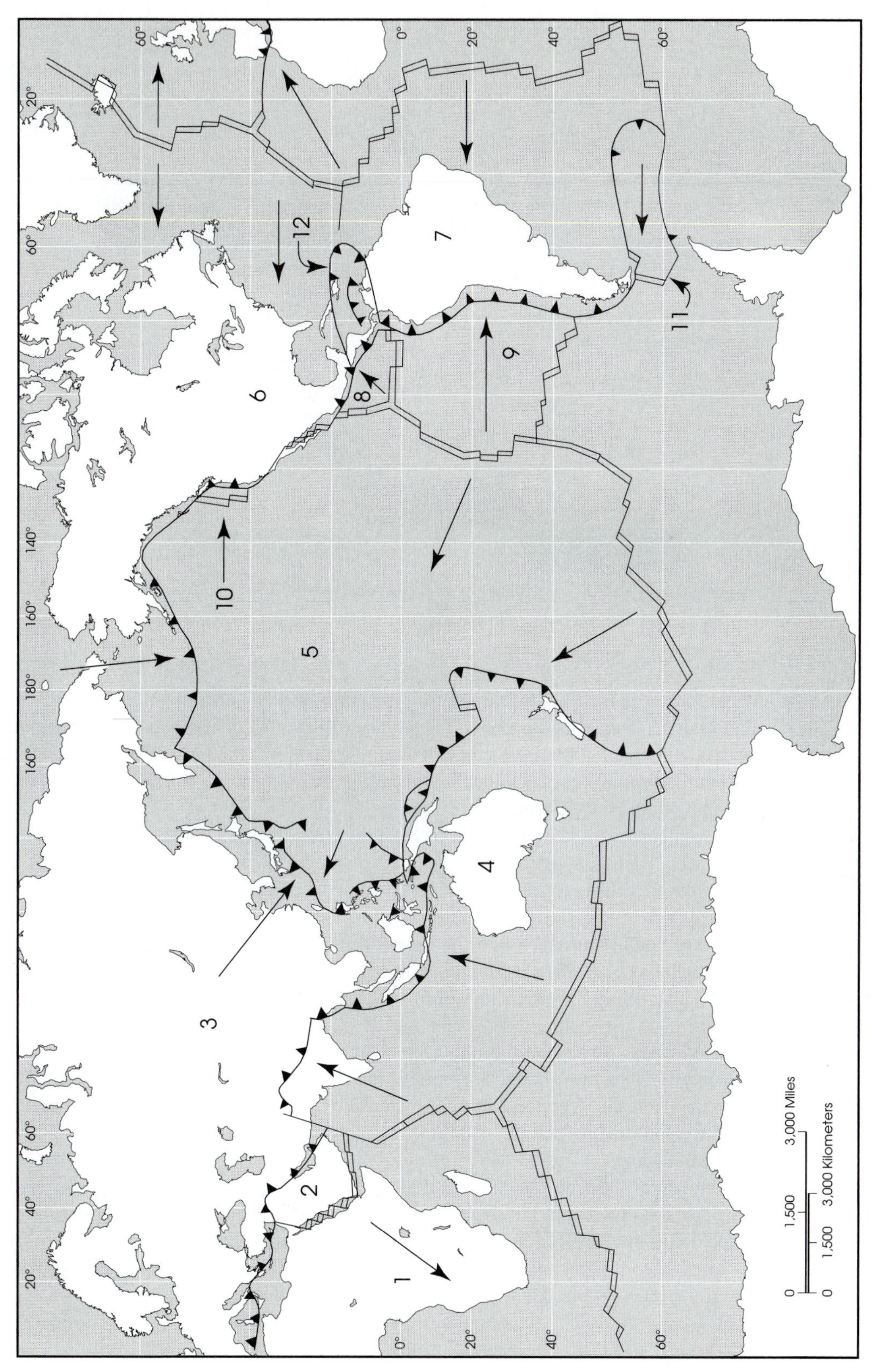


**Figure 2.3** Major plates making up Earth's crust. Each of the continents lies on a separate plate: 1. African plate; 2. Arabian plate; 3. Eurasian plate; 4. Australian plate; 5. Pacific plate; 6. North American plate; 7. South American plate; 8. Cocos plate; 9. Nazca plate; 10. Juan de Fuca plate; 11. Scotia plate; 12. Caribbean plate.

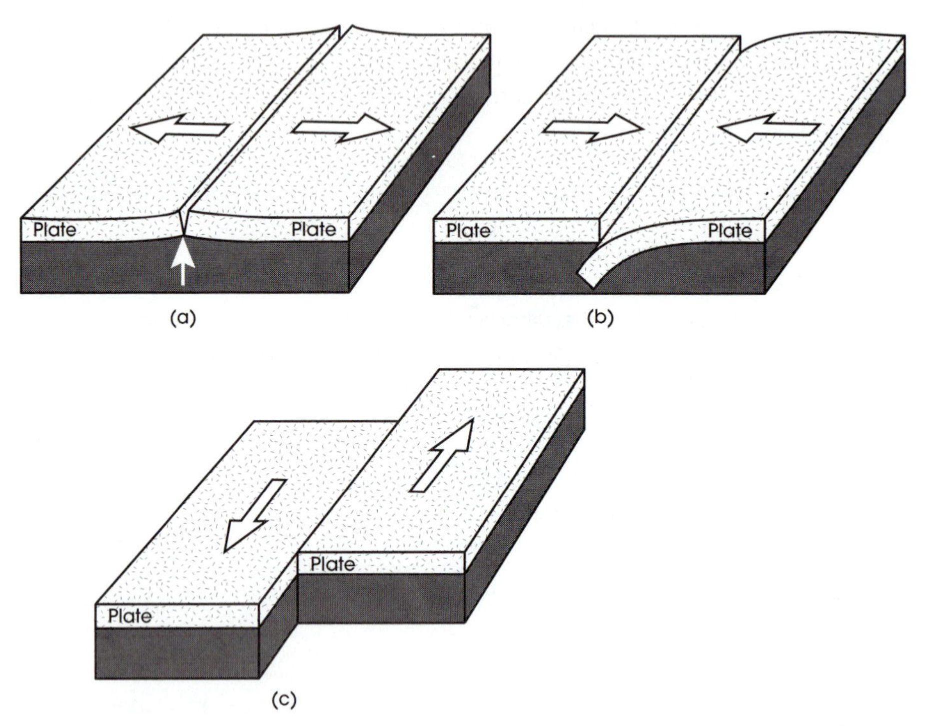


**Figure 2.4** Types of plate boundaries: (A) divergent boundary; (B) convergent boundary; (C) transform boundary.

California has become fused to the Pacific plate and is moving northwest. San Francisco lies on the San Andreas fault, which is one of the faults along this boundary. The famous San Francisco earthquake and fire of 1906 was a result of a rupture along this fault line. The earthquake of 1989 in northern California was on another fault, near the San Andreas fault. There are many such fault lines running through California.

If two plates move toward each other, a convergent boundary forms. In this case one plate moves beneath the other (subduction). The friction produced by pushing one plate down into the mantle produces melting of the rock and upheaval of the adjacent crustal plate. This builds mountain ranges such as the Cascade Mountains of the northwestern United States and the Andes Mountains of South America. If two continental plates converge, major uplifting takes place. The highest mountains on Earth are the Himalaya Mountains in Asia. These mountains and the high Tibetan Plateau resulted from the collision of the plate beneath the Indian subcontinent and the Asian plate.

Where plates are separating from each other, new crust forms at what is known as a divergent boundary. Along these boundaries melted rock from within the mantle emerges as lava flows or as ejecta in volcanic eruptions. The Mid-Atlantic Ridge is an area of active seafloor spreading of which Iceland is a product. The rift runs through the island and there are active volcanoes and frequent earthquakes.

Continental plates are also torn apart along divergent boundaries. From Syria south through Africa a crack is opening in the crust. The Dead Sea and the ancient city of Jericho lie near the north end of this rift. The Red Sea is along it, and in Africa it is known as the Rift Valley. Throughout its length there is heat near the surface. There are small active volcanoes, extinct volcanoes, geysers, fumaroles, and frequent earthquakes.

As the plates move about over Earth, new crust forms along divergent boundaries and old crust is destroyed along convergent boundaries. These processes create mountain ranges and deep-sea trenches. These processes also create the earthquakes and volcanic eruptions that occur daily on the planet.

Evidence of plate tectonics goes back more than 2 billion years in both North America and Asia. The core of the North American continent fused together between 1.98 and 1.83 billion years ago. Paul Hoffman of the Geological Survey of Canada pieced together the formation of the continent from smaller pieces of crust he calls microcontinents. The continental core included Greenland and part of Scandinavia. Collision with other continents added chunks to North America until the end of the Precambrian. The oldest rocks on the continent are in the interior and date from 1 to 2.5 billion years ago. The mountain regions of the eastern and western coastal regions range mostly from 200 to 500 million years in age.

Natural diamonds develop under a set of extreme heat and pressure conditions that occurred only as the crust formed. All natural diamonds are more than 2 billion years old. Since no new continental crust is forming, there are no new deposits of diamonds. Plate tectonics theory also explains why diamonds exist in South Africa and Australia in large numbers. These areas have some of the oldest pieces of crust still left.

## THE PRIMITIVE ATMOSPHERE

As Earth cooled and a solid crust formed, gases such as carbon dioxide, nitrogen, and water vapor formed. These gases made up the secondary primitive atmosphere. It may have had a composition of 60 to 70 percent water vapor, 10 to 15 percent carbon dioxide, and 8 to 10 percent nitrogen. Sulfur compounds made up the remainder. In this early environment radioactive decay of potassium-40 in the crust added the gas argon to the atmosphere. This mix of gases was 10 to 20 times more dense than now. The high concentration of carbon dioxide trapped much of the radiation given off by the Earth. Temperatures near the surface were much higher then than now; they were in the range 85 to 110°C (185 to 230°F).

By 3.8 billion years ago continued cooling caused the water vapor to condense, clouds to form, and frequent, intense, and widespread rain to occur. Initially, the rain did not reach the surface. At the high temperatures that still existed at the surface, the water evaporated well above the surface. Eventually, the surface temperature cooled enough to allow rain to reach the surface. The hydrologic cycle began to operate as water changed from liquid in the ocean, to gas in the atmosphere, and back to liquid precipitation. The movement of water from sea to land began and lakes and rivers became part of the landscape. The large amount of water vapor in the atmosphere must have caused rains that went on continuously for thousands of years. The rain washed out large amounts of carbon dioxide, which then chemically combined with surface materials to form carbonate rocks.

## POSSIBLE ORIGINS OF LIFE

The most important event of the Precambrian is the appearance of living organisms. Exactly when or where life first appeared on the planet is not known. Several aspects of the appearance of life are certain. All the chemical elements essential to life were present before life appeared and are present in seawater. The volume of water in the world ocean has been near present levels for at least 3.5 billion years.

A. I. Oparin, a Russian scientist, proposed in the 1930s that life began in a "primordial soup." He assumed that the early atmosphere contained all the necessary ingredients for life. There were carbon compounds such as amino acids, carbohydrates, and sugars in solution in atmospheric water, in addition to ammonia, methane, and nitrogen. He suggested that either solar radiation or lightning triggered chemical reactions. These reactions produced complex organic molecules that could reproduce. Another alternative is that the earliest organisms developed in clay. Clays often act as catalysts in chemical reactions. They can also absorb and transfer energy. The energy stored in clay may have triggered the chemical reactions producing organic compounds.

## RADIOMETRIC DATING OF EARLY EVENTS

Throughout most of history scientists had to settle for relative dating of geological events. Relative dating sets the time of an event or the age of rock by its relative position with respect to other events or other rock layers. There is no accurate way of determining how many years ago an event took place. In the past century a variety of discoveries and technological advances have provided methods of dating events and the age of rock. Radiometric techniques provide a reliable means of dating events in the Precambrian. There are five primary radiometric clocks in use (Table 2.4). One of these uses uranium isotopes 238 and 235. These isotopes decay into the lead isotopes 206 and 207. Decomposition of uranium isotopes into lead takes place at a known rate. Therefore, the ratio of uranium to lead provides the age of rock. Crystals of zircon often contain uranium isotopes. These crystals form when granitic magma solidifies. The use of uranium isotopes thus is particularly important in dating ancient lava deposits. Uranium-238 is most accurate for dating rocks of 4.5 billion years in age. Uranium-235 is most accurate for dating rocks of 713 million years in age. These two isotopes thus work well in dating rocks that date from the Precambrian Era.

Primitive life-forms developed in the period from $3\frac{1}{2}$ to 4 billion years ago. Carbon isotopes also help determine if fossil substances represent living organisms. When living organisms manufacture protein, they incorporate carbon in the form of several stable isotopes that have a small variation in mass. The most common isotopes of carbon are carbon-12 and carbon-13. In Earth's environment there is more carbon-12 than carbon-13 by the ratio of 90:1. Plants, however, store carbon-12 in a little different ratio. If the ratio of carbon-12 to carbon-13 in remnants of organic matter differs from the environmental ratio, it is an indication that the residue represents the remains of living organisms. Scientists working on Greenland have recovered tar from sediment that is 3.8 billion years old. The carbon-12/carbon-14 ratio shows that the tar came from living organisms. These organisms were extremely small and without any hard skeleton that could readily leave fossil imprints.

**TABLE 2.4** Isotopes Frequently Used In Radiometric Dating

| *Radioactive parent* | *Stable daughter product* | *Currently accepted half-life values* |
|---|---|---|
| Uranium-238 | Lead-206 | 4.5 billion years |
| Uranium-235 | Lead-207 | 713 million years |
| Thorium-232 | Lead-208 | 14.1 billion years |
| Rubidium-87 | Strontium-87 | 47.0 billion years |
| Potassium-40 | Argon-40 | 1.3 billion years |

The oldest fossils yet found date from about 3.5 billion years ago. J. William Schopf, director of UCLA's Center for the Study of Evolution and the Origin of Life, found the fossils. They were in some well-preserved Archaen rocks in Australia called the Warrawoona. The fossils are of bacterialike microorganisms. There are also several varieties. They are extremely small, being only about 2.5 thousandths of a centimeter in length. These bacteria do not require free oxygen to live. In fact, they lived in a carbon dioxide–rich environment that was largely free of oxygen. They were the predominate form of life for 2 billion years.

Oxygen did not begin to build up in the atmosphere until long after the planet formed. The first free oxygen probably resulted when water vapor dissociated to form hydrogen ($H_2$) and oxygen ($O_2$). Early levels of atmospheric oxygen were only 0.1 percent that of the present. Gradually, the amount of oxygen increased. The lighter hydrogen escaped to space. Traces of oxygen remained to begin formation of an ozone ($O_3$) layer some 30 kilometers above the surface.

Energy from the sun is emitted in various wavelengths, including the ultraviolet range. In large doses ultraviolet radiation is lethal to most forms of life found on Earth. Ozone plays a basic role in screening shortwave ultraviolet radiation from reaching Earth's surface. It absorbs it high in the atmosphere. The accumulation of atmospheric oxygen was an important element in the evolution of life on the planet. When the level of oxygen reached 1 percent that of the present, it screened enough ultraviolet radiation to permit sustained life in the oceans.

## *PHOTOSYNTHESIS AND THE ACCUMULATION OF OXYGEN*

The next step in the process of evolution was the development of organisms capable of photosynthesis. Photosynthesis removes carbon dioxide from the atmosphere. The process takes carbon dioxide and water, and in the presence of sunlight, produces oxygen. Between 3.5 and 4 billion years ago crude forms of blue-green algae appeared. They were oxygen-producing bacteria that lived in an oxygen-poor environment. An example of these organisms is the stromatolite, the most common fossil from the Precambrian. These fossils are layered mounds of minerals laid down by primitive microorganisms. They appeared about 3.5 billion years ago, increased in numbers up to about 1 billion years ago, and then began to decline. They are marine organisms restricted to shallow water. They need light to carry out photosynthesis, but they do not tolerate much ultraviolet radiation. Initially, they could not live on the surface. They lived in a zone below the surface where there was sunlight but little ultraviolet light. Fossil remains of these photosynthetic bacteria exist in Western Australia. Living stromatolites that are made up of mats of algae still live in Hamelin Pool in Western Australia.

Blue-green algae grew in numbers and supplied oxygen to the world ocean. The early ocean contained a large amount of dissolved metal. Once photosynthesis began in the ocean, the oxygen combined with iron and other metals to form oxides and hydroxides. Marine sediments dating from 3.8 billion years ago contain red-colored formations of iron compounds. Formation of oxides removed most of the oxygen produced in the sea for the next 2 billion years. The quantity of iron in the ocean and the slow rate of oxygen production made the removal of the dissolved metals a long process.

It was not possible for nonbacterial forms of life to develop until there was a surplus of oxygen ($O_2$) in the ocean. The dominate life-forms of the middle Precambrian were aerobic prokaryotes. *Aerobic* implies a need for oxygen or a tolerance for it. The term *prokaryotes* designates the most basic level of cellular organization.

**Figure 2.5** Inner gorge of the Grand Canyon. The Colorado River has cut deeply enough into the Colorado Plateau to reach precambrian rocks. The upper part of the canyon is being widened by erosion by wind and water. The river is now cutting an inner gorge into rock some two billion years old.

## *MASS EXTINCTIONS*

Global or regional environmental change can cause the extinction of living organisms. Evolutionary adaptation and ecological equilibrium cannot take place as rapidly as the environment can change. Regional or global environments can change in days, months, or years. Adaptation to environmental change by plant and animal species may take years, centuries, or thousands of years.

Plants and animals respond at different rates and in different forms to environmental change. Plants are always producing hybrids and individuals with mutations. In any plant species, there are those individuals better suited to the extremes of the range of the plant. These individuals may thrive under changing conditions while the majority perish. In this manner, the optimum conditions for the species changes to those of the new environment.

Animals, however, gradually come under more and more stress with changing environmental conditions until they die out. At intervals through geologic time events took place that resulted in the demise of most living species. There have been at least 13 mass extinctions since life became abundant on the planet. Five of these represent the most severe losses. As a result of these mass extinctions 99 percent of all species that existed on the planet no longer exist.

The first mass extinction took place over a long span of time. It occurred when oxygen replaced carbon dioxide as the primary gas in the atmosphere. Organisms called eukaryotes appeared about 1.5 billion years ago. They are single-celled organisms with a nucleus and organelles. More complex than prokaryotes, they represented a major biological advancement. These oxygen producers began to change the atmosphere from carbon rich to oxygen rich. Existing organisms had to adapt to the change or die. Most organisms succumbed to the change, which resulted in the first mass extinction. This change from a carbon dioxide–rich to an oxygen-rich atmosphere killed nearly all forms of organisms that existed before the change.

As the level of oxygen in the atmosphere increased, the level of carbon dioxide decreased. Carbon dioxide is soluble in seawater, and the ocean absorbed a lot of the atmospheric carbon dioxide. Carbon dioxide is a major ingredient in carbonates that make up the shells of marine organisms. As the numbers of organisms with shells increased, they took an increasing amount of carbon dioxide from the atmosphere. When the organisms died the shells sunk to the bottom, forming carbonate-rich sediments.

When the oxygen content of the atmosphere reached 10 percent of the present content, it protected the land surfaces as well as the oceans from excessive ultraviolet radiation. Life soon appeared on land. Primitive land plants appeared at the end of the Silurian Period (408 million years ago). By the Carboniferous Period (360 million years ago), forests were widespread over the Earth.

Gradually, the oxygen content of the atmosphere increased to present levels. In fact, it is possible that there have been times in the past when the oxygen content was far greater than it is now. The present atmosphere is one in which nitrogen and oxygen are dominate, making up about 99 percent of the gaseous mix.

## *GLOBAL CLIMATIC CHANGES*

Throughout most of the history of Earth, the planet has been much warmer than it is today. At the beginning of this chapter we mentioned that ice ages occurred during the Precambrian. The earliest ice age took place 2 billion years ago. The evidence for this glaciation is largely in Canada.

The second ice age took place from 800 to 600 million years ago and was more widespread than the previous one. Ice collected near the polar regions and expanded outward. The glaciers repeatedly scoured the continents. Evidence of this glacial period comes from Greenland, Scandinavia, Africa, Australia, and eastern Asia. This may have been the most extensive glaciation ever to occur on the planet. This glaciation also correlates with the boundary between the Precambrian Era and Paleozoic Eras. Table 2.5 provides approximate dates for the major ice ages. Dates of glaciation vary for different parts of the Earth because of regional differences in climate. The dates also vary depending on the dating methods used.

There were major changes in sea level taking place near the boundary between the Precambrian and the Paleozoic. Africa, South America, Australia, and the Antarctic were one large supercontinent until near the end of the Precambrian. The landmass often referred to as Prot-Pangea began to split up as seafloor spreading became pronounced. Ridges formed through seafloor spreading and displaced large volumes of water, causing sea level to rise. The rising seas flooded large areas of coastal lowlands. This provided a large area of shallow-water habitat that undoubtedly played a part in the development of hard-shelled marine animals.

**TABLE 2.5** Major Planetary Ice Ages

| *Approximate age (millions of years ago)* | *Event* |
|---|---|
| 0–1.8 | Pleistocene |
| 250–350 | Permo-Carboniferous |
| 450 | Ordovician |
| 650 | Early Cambrian |
| 750 | Late Precambrian |
| 950 | Early Precambrian |
| 2250 | Archeozoic |

One set of theories of climatic change depends on the location of continental landmasses in relation to the poles and the equator. Reconstructing the location of the continents during the last two global cool periods results in an interesting pattern. Maps of the Permo-Carboniferous glaciation (250 million years ago) and the Pleistocene glaciation show landmasses near one or both poles. Landmasses in polar areas are much more conducive to glacial formation because they lose heat rapidly in winter. It may be that a primary requirement for the formation of ice caps on continents is a polar location. However, there have been times when continents were near the poles but no glaciation resulted. A polar location of the continents may be necessary for ice caps, but actual ice formation must rest upon other causal factors.

Mountain building and continental uplift are part of the theory of climatic change. As an example of this theory one can use the formation of permanent snow on Mount Kilimanjaro. The mountain is in Tanzania, astride the equator. As the height of landmasses increases, the potential for snow and ice formation and accumulation increases.

Geologists have long noted the relationship between times of extensive mountain building and ice ages. For example, both the Permo-Carboniferous and recent ice ages followed extensive mountain-building periods. There is resistance to this theory from two directions. First, there is a gap of millions of years between mountain-building and the onset of glaciation. Second, some mountain-building periods—for example, the Caledonian Period in Europe (370 to 450 million years ago)—did not cause ice ages.

Despite these criticisms, mountain building, or at least the presence of high mountains, certainly adds to favorable conditions for ice formation. Modern research shows that mountain ranges influence upper-air circulation patterns. When sea level drops during glaciation there is a relative increase in the heights of the continents and landmasses increase in area.

Variations in sea-surface temperatures are linked to changing weather and global circulation patterns. Sea temperatures vary because of changes in salinity, evaporation rates, and the amount of solar radiation reaching the surface. Ocean currents play a significant role in moving energy over Earth's surface. Ocean currents transport large amounts of heat, and any changes in their relative extent and location has major implications for global climate.

There is no single theory that can account for all changes in global climate. It is clear that Earth's climate results from many causal elements.

## *SUMMARY*

The Precambrian is the longest of the geological eras. It encompasses 80 percent of the history of Earth. During this long span of time Earth changed drastically. Earth began as the result of an accumulation of debris circling the sun. In the beginning it was a hot molten mass without atmosphere, ocean, or land. The planet continued to receive strikes of large and small pieces of debris. It was only after a long time that the planet cooled enough for oceanic crust to form. Continental crust formed still later.

Initially, the planet was extremely hot. As it cooled, water condensed to form clouds high in the atmosphere and eventually resulted in torrential rains, which formed our ocean. The initial atmosphere contained high concentrations of carbon dioxide and little oxygen. Eventually, the balance changed to that of the present with some 20 percent oxygen. During this long epic of change, temperatures varied from hot to cold. The first ice age took place in this early time.

Life-forms appeared during this early time only to face extinction from disasters such as bolide impacts and slowly changing global conditions. A major mass extinction occurred when the atmosphere changed from carbon rich to oxygen rich.

## BIBLIOGRAPHY

DALZIEL, I. W. D. 1995. Earth before Pangea. *Scientific American,* 272:58–63.

DONOVAN, S. K., ED. 1989. *Mass Extinctions.* New York: Columbia University Press.

LANE, G. N. 1993. *Life of the Past.* Englewood Cliffs, N.J.: Prentice-Hall.

LEVIN, H. L. 1994. *The Earth through Time.* 4th ed. New York: Harcourt Brace.

# chapter 3

# Late Geologic Time

***CHAPTER SUMMARY***



Precambrian time spanned some 80 percent of the early existence of the planet. The Paleozoic, Mesozoic, and Cenozoic Eras encompass the remainder of geologic time. More evidence, and a greater variety of evidence, is available about the environment during these eras.

Each boundary between eras and periods in the geologic timetable represents a time of unusual change. The Precambrian–Cambrian boundary is an extremely important time of change in earth history. The life forms that existed at the end of the Precambrian included bacteria, blue-green algae, corals, jellyfish, worms, and seaweed. None of these organisms had a shell or skeleton. It is for this reason that fossil remains are few. Nearly 90 percent of the total history of the planet passed before any living organisms appeared with skeletal parts. It is the development of animals with skeletons and the explosion of species that distinguishes the early Paleozoic Era.

## *EARLY CAMBRIAN LIFE*

The Cambrian Period marks the time in earth history when the first hard-shelled organisms appear in the fossil record. The first skeletons were external. Snails and clams are examples of these. There are very important reasons for skeletons developing in animals. Life of the early Paleozoic Era consisted of marine life adapted to shallow water. Most marine life is in the upper 100 meters, where light penetrates and green plants can live. An outer skeleton provided protection from ultraviolet radiation in shallow waters, or near the surface of the open ocean. The outer skeleton offered some protection from predators. Skeletons also provided a frame upon which to attach strong muscles. This increased their ability to move around.

Five hundred fifty million years ago, trilobites appeared. They had an external skeleton made of chilton, a hard calcium-rich substance, and they are the first species known to have developed eyes. Trilobites were mobile, either crawling on the bottom if they developed a thick shell, or swimming if they had a lightweight shell. They were extremely successful; they could adapt to different marine environments. Trilobites became worldwide in distribution and were the most abundant living organism of the Paleozoic. Trilobites make up 60 percent of all early Paleozoic fossils.

Trilobites are part of the group of arthopods, all of which are invertebrates that lack backbones. Crabs and lobsters are arthopods that exist today. In addition to the trilobites, the brachiopods and mollusks developed early. All the major forms of invertebrate animals had

**Figure 3.1** Zion Canyon, Utah. The canyon has been cut during the past few million years following the tectonic uplift of the Colorado Plateau. The rocks into which the canyon is cut are Paleozoic sediments.

developed by 500 million years ago. After that, diversity of species in the ocean increased very rapidly.

It is bone, or an internal skeleton, that is the identifying characteristic of the vertebrates. The first fossil evidence of the vertebrates comes from early Paleozoic time. The first vertebrates probably developed early in the Cambrian, but evidence is not yet conclusive. However, the oldest vertebrates that are agreed upon do not appear until the Ordivician about 460 million years ago. These early vertebrates were small, fishlike animals that developed in the sea. Eventually, they colonized brackish water and fresh water. They all became extinct within 100 million years, but they were the forerunner of all vertebrates. Like the trilobites, they had a bony external covering. These fish were scavengers of the mud on the seafloor and lacked jaws or teeth. They developed nervous systems and sensory organs such as a sense of smell. As a result, they soon developed advantages over other organisms. They could search for prey; they could capture the prey they found; and they could escape their predators.

## *LIFE MOVES TO THE LAND*

Land-dwelling plants appeared in the Silurian Period, or middle Paleozoic. The first successful plants to live on land were probably green algae. The change from living in the sea to living on land must have been difficult. The marine environment is far more stable than the terrestrial environment. The problems fall into two groups. The first is how to handle the extremes of temperature, wind, moisture, and other variable elements of the continental environment. The second is how to handle the lack of water, or a reduced amount of water. They did this by developing a stronger and thicker skin, both for support and for reducing water loss. The oldest known land plants are the psilophytes, very simple plants without true leaves or root systems. They were spore-bearing plants.

Gradually, seed-bearing plants, or gymnosperms, appeared. They are in the fossil record by 400 million years ago. Seeds have an advantage over spores in that the embryo plant has an initial supply of nutrients and a protective case. Once they appeared they began to replace the earlier spore-bearing plants. The conifers of today appeared in the fossil record 300 million years ago. By the end of the Paleozoic the seed-bearing plants were dominate on the landmasses.

Animals followed the plants onto the landmasses only after a long interval. The first to go on land were the amphibians. These animals could live and feed out of the water but had to return to the water to lay their eggs. The first amphibians descended from the lobe-finned fish and the lungfish. The lobe-finned fish has pairs of muscular fins. They could use these fins to pull themselves across the ground when pools began to dry up. As they spent more time on land, their lungs, legs, and skeletons developed. The oldest known species, protoanthracosaur A, which dates back to 365 million years ago, were extinct by 300 million years ago. The oldest fossil found in North America comes from a quarry near Delta, Iowa. The species, which lived in subtropical lakes, resembles a crocodile or an alligator. It was 4 feet long, with large jaws, large teeth, and short legs.

Reptiles, which followed the amphibians, have the advantage of laying eggs. By laying their eggs on land, reptiles could finally free themselves from the sea.

## *FOSSIL FAUNA AS CLIMATE INDICATORS*

Vertebrate fossils provide important clues to past climates. Their distribution and physiology is related to the environment. The physiology of vertebrates is probably the most widely used method for interpreting the ecologic conditions in which they lived.

Fossil organisms and rock types suggest that the Cambrian and early Ordovician were largely warm. There may have been a glaciation toward the end of the Ordovician. The retreat of the ice and deglaciation at the end of this event resulted in climates in the Silurian and Devonian that were similar to those of today.

Following the ice age at the end of the Precambrian, the earth rapidly warmed. For the remainder of the history of the earth, temperatures have averaged 5°C higher than at the present. These warmer conditions existed probably 90 percent of the time over the past 570 million years. There are a variety of forms of evidence to support the higher temperatures. Among these are soil formations, rock types, cave deposits, reef formations, and plant and animal remains. Relict examples of these are now found in subpolar and polar regions.

In warm environments, particularly in warm wet environments, natural weathering processes produce soils highly enriched with aluminum oxide ($Al_2O_3$) and iron oxide ($Fe_2O_3$). These compounds stain the soil red. In extreme cases soils can consist of 75 percent $Al_2O_3$ and $Fe_2O_3$. The existence of fossil soils high in aluminum and iron oxides in North America and Europe indicates temperatures above those of the present.

Certain cave deposits show the temperatures in which they formed. Calcium carbonate ($CaCO_3$) crystalizes in one of two forms. It is deposited as aragonite at temperatures above 16°C and as calcite at lower temperatures. This is not an altogether reliable process, as exceptions do occur, depending on the chemistry of the cave itself. The crystal form of the $CaCO_3$ content of limestones is a sign of the water temperature in which they formed. Calcium carbonate is less soluble in warm water than in cold. Thus limestones derived from warm oceans have a higher calcite content. Many extensive limestone deposits of midlatitudes have high levels of calcite. Coral reefs now grow between 30°N and 30°S.

## *FOSSIL FLORA AS CLIMATIC INDICATORS*

Today's world plant distribution provides an important guide to the distribution of climate. The same is true of paleoclimates. Identification of vegetation patterns and their changes over time is widely used to interpret past climates. Often, the evidence is used in relation to other environmental features. The physiology of plants, like that of animals, provides much information. Drip tips on plant leaves show that they developed under very moist conditions. Plants with thick, fleshy leaves are characteristic of arid or semiarid climates.

Many of the fossil plants living in the Carboniferous were related to the horsetail and club mosses, both representative of a marsh or swamp environment. Other plant fossils show layered roots such as those found in modern bog plants. Other physiologic characteristics show that some of the plants actually floated on water. Trees of the period show a lack of development of growth rings, indicative of a climate without marked seasonal differences. In all, the representative vegetation suggests a warm, moist climate that favored a luxurious plant cover.

In today's energy-conscious world, we know the Paleozoic Era for the deposits of coal and petroleum that formed at this time. The Carboniferous period (360 to 286 million years ago) was dominated by widespread warm humid climates with intermittent glaciation. As the southern hemisphere ice sheet expanded and contracted, large changes in sea level took place. During periods of glacial retreat, sea level rose, drowning vast areas of tropical forests and burying them in ocean sediments. These buried layers of ancient tropical forest are the coal and oil deposits of today (Figure 3.2). Microorganisms buried in sediments were the raw materials of petroleum. In 1859 at Titusville, Pennsylvania, the first producing oil well was drilled. This well tapped a petroleum-saturated Devonian sandstone that was only 10 meters below the surface.

**Figure 3.2** The Carboniferous Period of the Paleozoic Era resulted in the formation of major deposits of coal and oil in North America. In this photograph, loaded coal trains are being assembled near Bluefield, Virginia.

The search for petroleum has provided most of the information we have about the crust of the Earth.

## *MASS EXTINCTIONS*

Several mass extinctions coincide with boundaries between geologic periods in the Paleozoic and Mesozoic Eras (Table 3.1). One mass extinction occurred in late Ordovician time. As in the previous mass extinctions, it destroyed mainly marine organisms. Entire families of species disappeared from the ocean. Another mass extinction coincided with the Devonian–Carboniferous boundary. Most of the world's fish and 70 percent of all animal species lacking skeletons perished.

The most extensive mass extinction of all time happened at the boundary between the Paleozoic and Mesozoic Eras. Ninety percent of all living species died out and very drastic changes took place in the fauna of the planet. There was a major collapse of marine fauna, especially those living in shallow water on the continental shelves. Among the organisms that perished were the trilobites. Trilobites were one of the dominant organisms of the seas. They survived the previous two extinctions, but could not withstand the changes that mark this period in earth's history. While many land species disppeared, there was no parallel to the mass extinction of marine animals. The changes were so drastic that the life of the following Mesozoic Era was quite different from that of the Paleozoic Era.

We do not know the cause of the extinction. Several environmental changes may have played a part. As the Permian ended, two landmasses, Gondwana and Laurasia, collided to form

**TABLE 3.1** Principal extinctions, probable causes, and primary losses

| *Extinction event* | *Probable cause* | *Primary losses* |
|---|---|---|
| Mid-Tertiary | Changes in sea level and ocean circulation. | Plankton, mollusks, mammals |
| Late Cretaceous | Disturbance by one or more bolide impacts. | Dinosaurs/ mollusks plankton |
| Late Triassic | Changes in sea level. | Marine invertebrates, some reptiles |
| Late Permian | Cooling associated with a long period of glaciation. Also sea level changes. | Marine invertebrates, reef builders, some reptiles |
| Late Devonian | Global cooling. | Plankton, marine invertebrates, primitive fish |
| Late Ordovician | Changes in glacial ice and sea level. | Marine invertebrates |
| Late Cambrian | Changes in sea level. | Trilobites Brachiopods Conodonts |

the massive supercontinent Pangea. The combining of these two landmasses altered climate and changed the pattern of ocean currents. As Pangea formed, large areas of shallow seas drained. This reduced area of shallow seas led to the demise of many species. This was also a time of unusual volcanic activity. Volcanism increases the acidity level of precipitation. This, in turn, changed the chemistry of the seas enough to bring about extinction in some species. The climate warmed during the early portion of the Paleozoic. Temperatures in the seas changed as well. With time the exact cause of the extinction may become clear.

## THE PERMO-CARBONIFEROUS ICE AGE

An ice age called the Permo-carboniferous began toward the end of the Paleozoic Era, about 325 million years ago, and lasted until about 250 million years ago. The south pole was in the midst of the large landmass called Gondwana (Figure 3.3). Ice sheets moved over about half of Gondwana. What is now Antarctica and parts of Australia, India, Africa, and South America were covered with ice. The glaciation of each of these areas did not take place at precisely the same time, but they were all affected by the same climatic cooling. The southern hemisphere suffered widespread glaciation, but the northern hemisphere remained warm. The most appealing explanation for this situation is a different relative location of the landmasses. The northern continents were nearer the equator and the southern landmasses nearer the poles.

## MESOZOIC CLIMATE

After the glaciation in the Permo-carboniferous ice age, the earth again entered a long period of warm conditions. The period of warmth continued through most of the Mesozoic Era. Throughout the Mesozoic, the earth was free of glaciation. Temperatures were warm and rainfall was

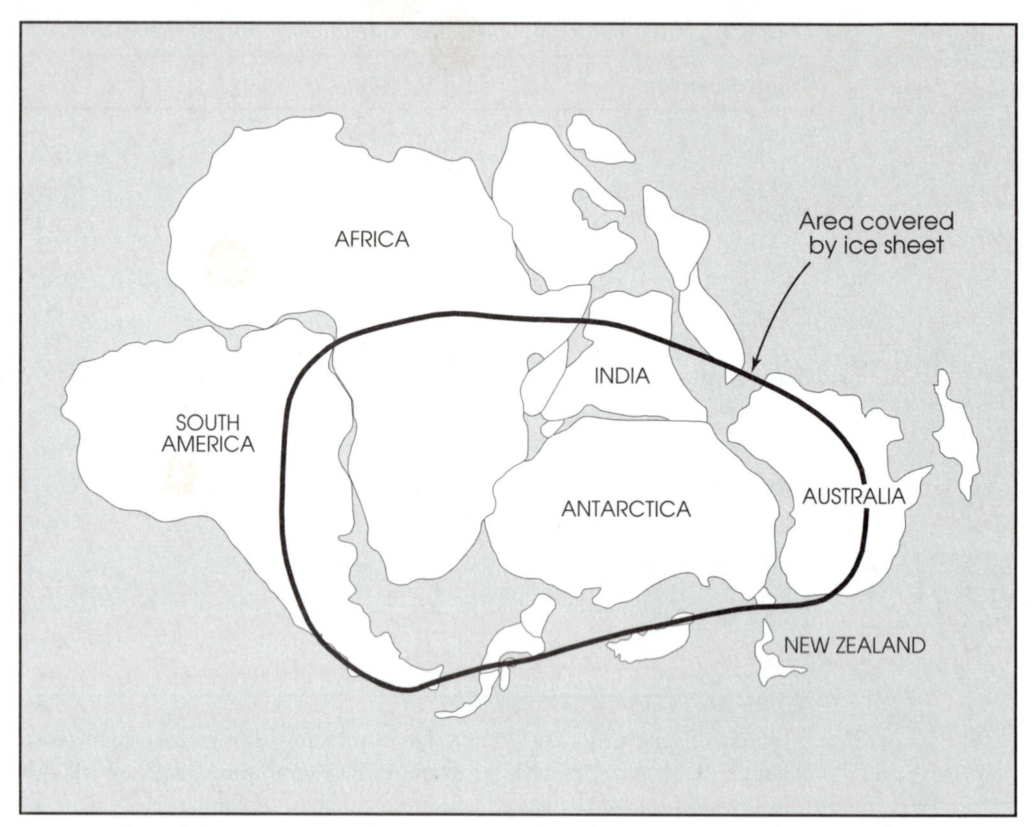


**Figure 3.3** Supercontinent Gondwana, showing the extent of the continental ice sheet (white area). Some 280 million years ago, ice covered over half of Australia and large sections of South America, Africa, and India.

abundant on the landmasses. Even the polar regions experienced mild weather. Initially, the warmer conditions resulted from the slow migrations of Pangea to the north. This carried areas that had been glaciated into warmer climates.

In the Cretaceous Period, Gondwana and Laurasia split, with the pieces moving toward their present positions. Once again, ocean currents changed. It warmed in the Cretaceous to a temperature that was the warmest time since the Precambrian. The climate at the south pole may have been subtropical. In early Cretaceous there was a global rise in sea level. The result was the flooding of large areas of land by shallow seas.

Major changes in the plant kingdom took place during the Mesozoic, especially so during the Middle Cretaceous. From this time forward the major plants have been part of the group known as angiosperms, or flowering plants. These plants pollinate through flower structures and bear seeds. Flowering plants inhabit all terrestrial environments and include herbs, shrubs, and trees. There are about 250,000 species today. Fossil evidence identifies another 30,000 species. By the end of the Jurassic Period early forms of terrestrial vegetation such as the conifers and ferns reached their peak.

Interdependent relationships between plants and animals developed along with flowering plants. The plants depend on the deposition of pollen containing male gametes to fertilize the seeds. Most flowering plants depend on insects to pollinate them. Most domesticated fruits, vegetables, and ornamental plants are among this group.

## *AGE OF REPTILES*

The Mesozoic is the age of reptiles. These animals became the dominant life-form during the era and remained dominate for nearly 125 million years. Turtles and crocodiles, as we know them, appeared early in the era. Snakes appeared toward the end of the era and were the last large group of reptiles to evolve. Reptiles inhabited the seas as well as the land. Sea turtles are the remaining survivors of the marine reptiles (Figure 3.4).

The dinosaurs, perhaps the best known inhabitants of the earth during the Mesozoic, developed during the Triassic Period. A variety of dinosaurs developed. Some were vegetarian, some carnivores, and some ate both meat and plants. Some walked on two legs and others were quadrupeds. Most were small, but some grew to very large size. Two orders developed, one being the lizard-hipped. They were carnivores and ran on their hind legs. As time went on, they developed very large size. *Tyrannosauris* weighed 5.5 metric tons and the head was 6 meters above the ground. They represent the largest carnivores to have lived on the planet. Later, a vegetarian form of the lizard-hipped dinosaur developed that eventually became the largest land animal to have lived on Earth. Some weighed as much as 36 metric tons and were nearly as long as a football field. A second group of dinosaurs was the bird-hipped species. They were smaller and were vegetarians. They probably formed a large part of the food supply of the carnivorous dinosaurs.

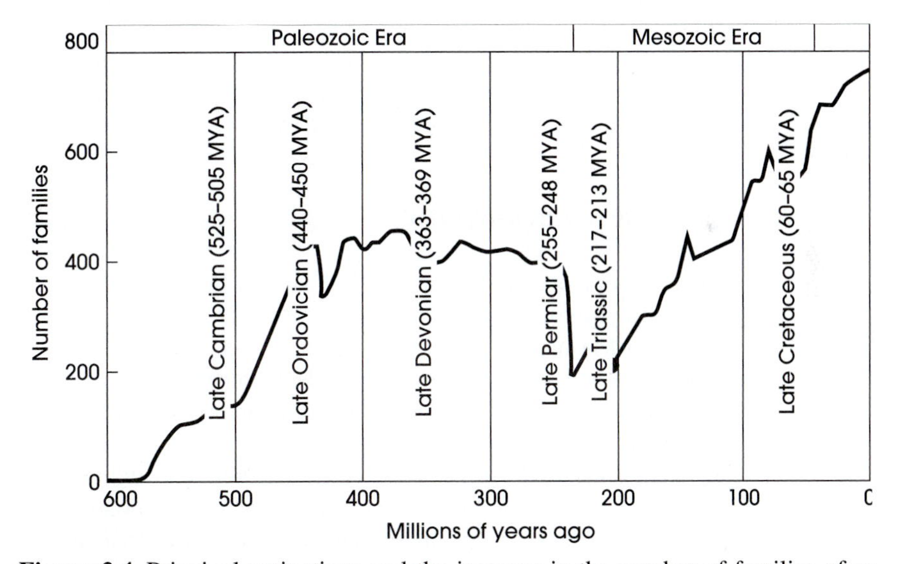


**Figure 3.4** Principal extinctions and the increase in the number of families of organisms during the Phanerozoic Eon. Dates for the extinctions and number of families are approximate. Data from Stephen K. Donavan (ed), 1989. *Mass Exctinctions*, Columbia University Press; Steven M. Stanley, 1987, *Extinction*. Scientific American Books; and E. O. Wilson, 1990. Threats to Biodiversity, in *Managing Planet Earth*. W. H. Freeman and Co.

Similar species lived on all the continents. This suggests that they were highly successful and that the landmasses were connected during the latter part of the Mesozoic.

Flying reptiles appeared in the middle of the Mesozoic. Most were small, but some reached very large size. One group, the pterodactyls, had very strong necks, powerful jaws full of teeth, and a large wingspan. Another, the pterandon, had a wingspan of at least 9 meters. One fossil specimen shows a wingspread of 12 meters. The pterandon is the largest known flying animal to inhabit the earth.

Another group of flying reptiles developed feathers. A fossil imprint from middle Mesozoic strata in Europe clearly shows feathers on the tail and wings. This imprint of archaeopteryx is the oldest known fossil of a bird. Modern birds descended from these animals, which were intermediate between the reptiles and the true birds. True birds have several distinctive features that make them good fliers. Among these features are hollow bones, extensive air sacs around the lungs, and four-chambered hearts. Feathers replaced scales over most of their bodies.

The Mesozoic Era (middle life) was witness to the rise of mammals. Mammals evolved 180 million years ago. The earliest forms were very small shrewlike animals that gave birth to live animals rather than laying eggs. A similar creature is the ancestor of all egg-laying mammals. The fossil jawbone of a close relative of these two animals was found in the Painted Desert of Arizona. The environment in which the creature lived was that of a swamp or a floodplain. The oldest fossil of a mammal found in Australia dates from 100 million years ago. It is an opalized jaw of *Steropodon galmani*, an early form of the platypus.

An important biological development took place in the seas during the latter part of the Cretaceous. There was an explosion of small floating organisms, both plants and animals. The plants were single-celled organisms. The quantity of microorganisms was so great that it reduced the carbon dioxide content of the atmosphere. The small animals had shells made of calcium carbonate ($CaCO_3$). The death of the small shelled animals resulted in a rain of shells to the seafloor. The layers of shells formed chalk deposits. These chalk deposits are now above sea level in many places over the earth, the white cliffs of Dover being a well-known example.

## MASS EXTINCTION AT THE K-T BOUNDARY

Another period of extinction marks the boundary between the Mesozoic and Cenozoic Eras some 65 million years ago. This boundary, known as the K-T boundary, marks the point at which Cretaceous rocks (mapped as K), adjoin Tertiary rocks (mapped as T). The extinctions were not instantaneous and were selective. Major environmental changes took place during the last few hundred thousand years of the Cretaceous Period. At the K-T boundary more than half of all plant and animal groups and 70 percent of all living species disappeared. Species living in the ocean and on the land died out. The event destroyed billions of living organisms.

In the oceans there was a sudden and drastic reduction in microscopic surface organisms. The number of these tiny plants and animals dropped almost to nothing and never again recovered to the levels of the Cretaceous. More than 90 percent of small plankton with shells became extinct. The plankton died from a lack of sunlight or a sudden increase in acidity of the water. Rain as acidic as battery acid increased the acidity of seawater and dissolved the calcium shells. The reduction in the small surface organisms lead to a reduction in all species higher up the food chain. Most living organisms disappeared from the surface layer of the ocean for more than 400,000 years.

The temperature of Earth's lower atmosphere rose by 5 to 10°C after the K-T event and stayed warm for tens of thousands of years. One group of scientists believes that the massive dying of small oceanic plants brought about this heating. Living microorganisms produce a cooling effect in the atmosphere. They produce a sulfur compound that enhances the development of clouds. Clouds, in turn, reflect sunlight and prevent radiation from reaching the earth. With fewer clouds more radiation reached the ground and heated the atmosphere.

Many land animal species ceased to exist. Nearly all land animals weighing more than 25 kilograms declined. The most publicized species that became extinct were the dinosaurs. They had developed, diversified, and persisted for over 150 million years. Fossils in the Painted Desert place dinosaurs in the area at least 225 million years ago. They lived in a wet and forested region and were plant eating.

Fossils of dinosaurs even exist near the Colville River on the north slope of Alaska. The site was at 70 to 85°N latitude at the time, some 600 to 1600 kilometers farther north than now. The presence of the bones of very young animals suggests that the dinosaurs did not migrate but stayed year round. It was a region of deciduous forest, so there was limited food in the winter. If they did stay through the winter, they could tolerate up to several months of cold weather, darkness, and limited food supply. Possibly, they entered a state of diminished activity to survive the winter.

The dinosaurs thrived during most of the Cretaceous Period but are missing from the Tertiary Period. They died out gradually, but some may well have survived the boundary event. Most mammals and birds survived. Of the land animals, the largest suffered the most but were not the only group to disappear. In all cases of severe environmental stress, it is the largest animals that suffer most. The event, or events, that brought about this massive elimination of organisms clearly produced major evolutionary changes.

## *CAUSES OF THE K-T MASS EXTINCTION*

There are many possible causes for the mass extinction. Two of these explanations suggest that the extinctions took place gradually. The first hypothesis is competition from species that were becoming better adapted to the environment. The success of the mammals perhaps crowded out other organisms. The second explanation suggested for a gradual dying out is that of a climatic change. Other possibilities involve rapid extinction resulting from short-lived events. Three hypotheses involve astronomical events. They are the explosion of a nearby supernova, the passing of a cloud of interstellar dust, and the impact of a massive meteorite or comet.

In 1979, Luis W. Alvarez and Walter Alvarez, then of Lawrence Berkeley Laboratory, hypothesized that it was the impact of an object from space that caused the extinction. They based their hypothesis upon the discovery of a rare element called iridium. Iridium is not common in the crust of the earth but is more abundant in meteorites. The researchers further suggested that the impact vaporized the foreign body, throwing iridium-rich dust into the atmosphere. This dust formed a cloud that blocked sunlight from reaching Earth for years. As the dust settled, it formed the clay layer rich in iridium that is found in many different parts of the Earth. Opponents of this hypothesis note that the iridium could have come from volcanic eruptions.

Other evidence also points to an impact. Bruce Baker and his co-workers at the U.S. Geological Survey in Denver found shock-altered grains of quartz at several sites that date from the K-T boundary. Such grains of quartz form during large meteorite impacts. While iridium exists in volcanic rock, shocked quartz grains do not. These grains form only under extremely high pressures such as would occur with a large impact.

In 1985, Wendy Wolbach, Roy Lewis, and Edward Anders at the University of Chicago found very fine soot in the K-T deposits. The presence of the soot suggests that fire storms were part of the catastrophic event. Heat produced by the impact started fires. These fires destroyed large areas of natural vegetation. They also removed much of the oxygen and placed large quantities of carbon dioxide in the atmosphere. Smoke produced by the fires blanketed the earth and prevented sunlight from reaching the surface. This caused temperatures to plummet and increased the stress on plants and animals. The fires and cold that followed would have destroyed much of the vegetation and hence the food supply for animals.

Another bit of evidence supporting the impact theory is the detection of a mineral called stishtovite in the K-T deposits. Stishtovite is a dense form of silica formed by extreme pressure. Scientists under the direction of N. M. Mchone found the mineral near Raton, New Mexico. Stishtovite occurs only at sites connected with impacts of extraterrestrial objects. Volcanic eruptions bring rare elements and minerals to the surface but do not produce stishtovite. It cannot survive very long in a volcanic environment as heating causes it to break down. Sustained temperatures as low as 300°C cause stishtovite to be changed into a less dense form of silica.

In 1989, Jeffrey Bada and Meixun Zhao of the Scripps Institute of Oceanography found other evidence of an extraterrestrial impact in the form of select amino acids that are rare on Earth. All plants and animals manufacture amino acids that combine in chains to form protein in living cells. Amino acids also form when chemicals such as methane and ammonia exist in an oxygen-free environment. This set of conditions exists in space, and a variety of ammino acids formed this way are found in meteorites. Bada and Zhao found two of the most common amino acids often associated with meteorites in sediments dating from the K-T boundary. There were some limitations in the study. It was based on sediments found at a single site in Denmark, and the concentrations of the two acids were far different from those typically found in meteorites.

## *PROBABLE IMPACT SITES*

The object producing the impact must have been at least 10 kilometers in diameter. An object this large must have left a crater 150 to 350 kilometers across. If so, where is it? The nature of the debris from the boundary suggests that the impact was in the ocean near a continent. The problem in finding the impact crater is that plate tectonics has buried a lot of crust. If the object did land on the seafloor near a continent, chances are that the crater no longer exists. The seafloor of that time has been pushed deep in the crust and remelted.

On the other hand, the shocked quartz implies an impact on land. Quartz is abundant on the continents but scarce on the seafloor. There is one large impact crater, in Manson, Iowa, that dates from 73 or 74 million years ago (plus or minus several million years). The crater is 32 kilometers (19 miles) in diameter and buried under 30 meters (100 feet) of glacial gravel. It is also possible that there were multiple impacts spread out over time. There are at least five craters that could be the right age. They date from 57 to 65 million years (±9 million years).

One very promising site that may mark the impact crater is located near the town of Chicxulub, Mexico, in the northern part of Yucatan Penninsula (Figure 3.5). Magnetic surveys, gravity surveys, and well logs indicate a large circular structure some 180 kilometers in diameter. In the structure are rocks containing andesite, which has a composition similar to tectites. Well logs show that rock in the structure is highly shattered, as would result from a large impact. There are also grains of shocked quartz in the rock. Radiometric dating gives the age of the structure at

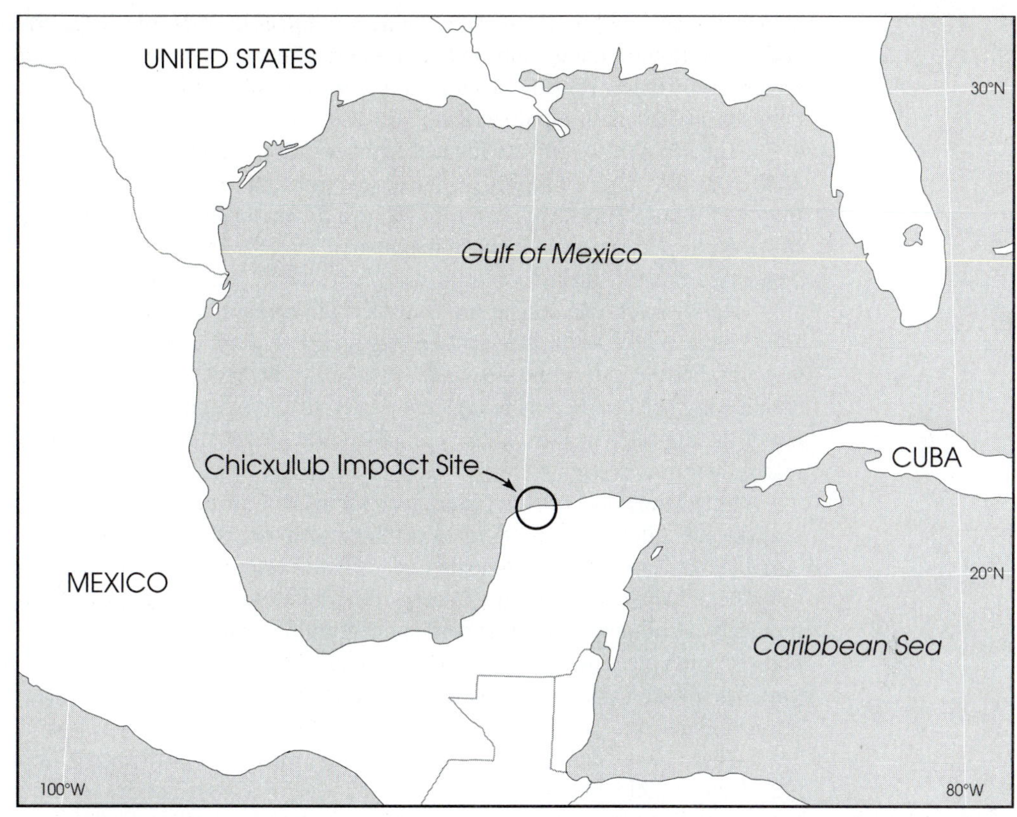


**Figure 3.5** Location of the Chicxulub structure in Mexico. The impact was in the center of the ring. The ring marks the location of sinkholes surrounding the site.

65.2 ± 0.4 million years. The rock also contain reverse geomagnetic polarity, which corresponds to the polarity at that time. The bolide that created the structure must have been at least 10 kilometers in diameter. The coastline around the Gulf of Mexico was struck by a wave of water that may have reached 34 kilometers in height. This is the largest impact crater yet discovered.

There are other potential impact sites as well. These include two sets of twin craters. One set is Kara and Ust-Kara in the arctic plains of the former Soviet Union. The other set is Gusev and Kamensk on the north shore of the Black Sea. Each pair occurred at the same time.

Alan Hildebrand and William Boynton of the University of Arizona considered the possible results of a meteorite impact in the ocean. They knew that if it hit in the ocean, it would generate huge waves. Other scientists calculated that a meteorite large enough to produce the iridium layer would produce a wave about 5 kilometers high. The impact of the object would scour the ocean floor over an area 1000 kilometers across. As the material scoured from the seafloor began to settle back, debris was deposited on the seafloor. They searched the scientific literature for references to deposits of course material. They found references to such sites in Central and North America and on the floor of the Caribbean Sea. Records show such deposits near Cuba that are some 450 meters thick. On North America there are some deposits 180 meters thick. They also found these deposits paralleled the continental margin as it was about 65 million years ago.

They found still other evidence. Just beneath the clay layer containing the iridium there is another distinct layer of sediment. This is a 19-millimeter-thick layer containing a large concentration of small tektites. On Haiti they found a 0.5-meter-thick layer of tektite-rich material. Some of the tektites are 64 millimeter in diameter. These are very large compared to the mean size of North American tektites.

Another possible impact site is in the Caribbean off the western tip of Cuba. One geologist found boulders 0.8 kilometers in diameter on the seafloor south of Cuba. He studied the distribution of these boulders and surmised that they must have been thrown to their locations rather than having been pushed there by tectonic activity. Other geologists suggest that Cuba is a product of the impact. The Isle of Pines in the central peaks and the curved tip of western Cuba represent the rim of the crater.

In 1974 members of the Geological Survey of Canada discovered a circular structure on the seafloor 200 kilometers (120 miles) southeast of Nova Scotia in 110 meters (350 feet) of water. An object at least a 1.5 kilometers (1 mile) in diameter produced the crater that is 50 kilometers (30 miles) in diameter. The crater dates from 50 million years ago, well after the event at the K-T boundary. Drilling revealed that in the middle of the structure there is rock that melted under sudden very high pressures, and also shocked quartz. The impact of a large object produced the structure. There is a more recent impact crater on Devon Island in Canada that is 20 kilometers (12 miles) in diameter. The Houghton Crater dates from 22.4 million years ago.

There have been several concentrated periods of impacts in recent geologic time (Table 3.2). It is possible that other impacts caused other mass extinctions. It is possible that large impacts caused most of the mass extinctions. Table 3.3 provides an estimate of the frequency of bolide impacts of various sizes.

We may never know for certain what caused this mass extinction. As evidence builds, there are two reasonable conclusions. One is that a collision of a large object from space is possible and probable. The second is that the extinction was complex and involved a variety of changes well beyond those of the blast impact. Coupled with other changes, an impact could well have added to the mass extinction. Cool, if not cold temperatures followed the impact for some months. Then temperatures rebounded to levels higher than those before the impact. This change in climate following the more sudden boundary event may well have completed the mass extinction.

## *SUMMARY*

The climate of Earth has oscillated widely over the last 570 million years. Vast ice sheets covered much of the planet at times. At other times Earth was warmer than it is now. Species of plants and animals diversified rapidly with time. Life on the planet was dealt a number of severe blows even as the number and complexity of species grew. Several mass extinctions of life took place during these eras. One of these destroyed nearly 90 percent of all living species. The best known of these extinctions, but not the most severe, is that at the end of the Cenozoic. This is the time of the demise of the dinosaurs. Debate over the cause of this mass extinction continues. It is quite probable that one or more collisions of large rock masses with Earth played a part in this catastrophe.

**TABLE 3.2** Major Times of Bolide Impacts Since the Precambrian

| *Period* | *Millions of years ago* |
|---|---|
| Pleistocene | 1 |
| Pliocene | 13 |
| Miocene | 25 |
| Oligocene | 36 |
| Eocene | 58 |
| Cretaceous | 65 |
| Jurassic | 181 |
| Triassic | 230 |
| Permian | 280 |
| Carboniferous | 345 |

**TABLE 3.3** Frequency of Impacts of Large Extraterrestrial Objects

| *Diameter (kilometers)* | *Frequency (millions of years)* |
|---|---|
| 3.9 | 9.2 |
| 5.0 | 14.0 |
| 10.9 | 58.0 |
| 30.8 | 360 |

## BIBLIOGRAPHY

BERNER, R. A. 1991. A model for atmospheric $CO_2$ over Phanerozoic time. *American Journal of Science*, 291:339–376.

BRADLEY, R. S., ED. 1991. *Global Changes of the Past.* Boulder, Colo.: University Corporation for Atmospheric Research (UCAR/Office for Interdisciplinary Earth Studies (OIES).

BRIGGS, D., and P. CROWTHER, EDS. 1990. *Paleobiology*. New York: Blackwell Scientific.

DONAVAN, S. K., ED. 1989. *Mass Extinctions.* New York: Columbia University Press.

ELDRIDGE, N. 1991. *Fossils.* New York: Harry N. Abrams.

FRAKES, L. A., J. E., FRANCIS, and J. I. SYKTUS 1992. *Climatic Modes of the Phanerozoic*. New York: Cambridge University Press.

STANLEY, S. M. 1987. *Extinction*. New York: Scientific American Books.

WILSON, E. O. 1990. Threats to Biodiversity. In *Managing Planet Earth*. New York: W. H. Freeman & Co.

# chapter 4

# The Pleistocene: The Age of Ice

***CHAPTER SUMMARY***



The last geological era, the Cenozoic, consists of two periods. The first, the Tertiary, encompasses a span of 63 million years. The second, the Quaternary, includes the last 1.8 million years of Earth history. The Quaternary Period contains two epochs, the Pleistocene and Holocene. The Quaternary is in many respects the most fascinating period of geologic time. During this time of rapid and extreme environmental change, the human species evolved into a position of dominance in the biological world.

## *ADVANCE AND RETREAT OF ICE SHEETS*

The most important single environmental event since the human species has been on Earth has been the oscillation between glaciation and interglacials during the Pleistocene Epoch. The epoch represents a large change from much of the last 600 million years. The Pleistocene Epoch represents the most recent of the several major cold periods to occur in the history of the planet. During the time when ice was most extensive over Earth, temperatures averaged 4°C less than at present. In the northern hemisphere it was perhaps 8 to 12°C lower than the current mean temperature.

By the beginning of the Pleistocene Epoch global cooling had begun. Major continental ice sheets formed in both hemispheres on landmasses near the poles. Mountain ranges supported massive glaciers that extended out into the surrounding lowlands and into the oceans. Pack ice covered the polar seas, and icebergs drifted far toward the equator in the cool ocean currents.

The Pleistocene ice age was not a single cool event followed by warming. It includes several large temperature changes. In the past 1 million years there have been many periods of glacial advance and retreat (Figure 4.1). Each period of advance includes many minor fluctuations. Fluctuations during the last million years have averaged about 100,000 years in length. Of the 100,000-year intervals, cold has dominated about 90 percent of the time. The interglacials, or warm stages, have been relatively short, somewhere in the range of 10,000 years. Even though the interglacials were relatively warm, temperatures still varied. The Sangamon interglacial stage, the last before the present, began about 130,000 years before the present. This interglacial lasted until about 110,000 years ago. Table 4.1 lists the North American glacial events. It is not possible to provide simple beginning and ending dates for the major stages of Pleistocene glaciation. The stages were long, uneven, and not everywhere correlated in time (Figure 4.2).

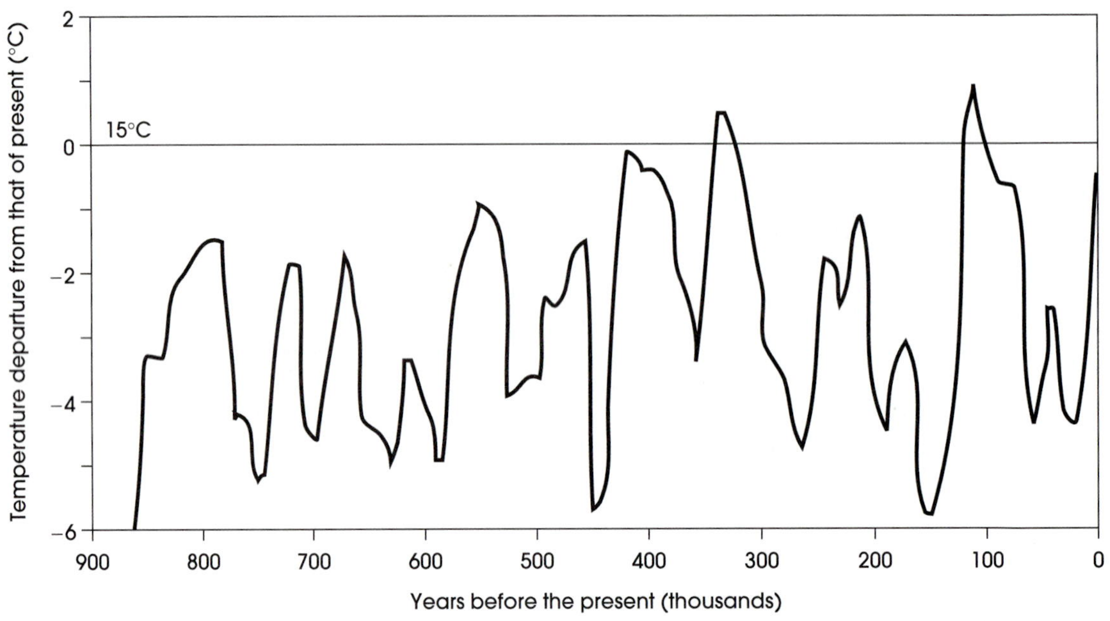


**Figure 4.1** Temperatures during the past 900 thousand years. The temperature record of the Pleistocene is being continually modified as more data become available. During the Wisconsin glacial advance, global temperatures may have been 4°C below those of today. During the Illinoian glacial advance it was even cooler. Global temperatures may have been 6°C below those of today.

**TABLE 4.1** Glacial and Interglacial Stages of the Pleistocene Epoch In North America

| *Estimated beginning of stage (thousands of years)* | *Glacial* | *Interglacial* |
|---|---|---|
| 100 | Wisconsin | |
| 128 | | Sangamon |
| 240 | Illinoisan | |
| 400 | | Yarmouth |
| 480 | | |
| 480 | Pre-Illinoian Glacials | |
| 550 | | |

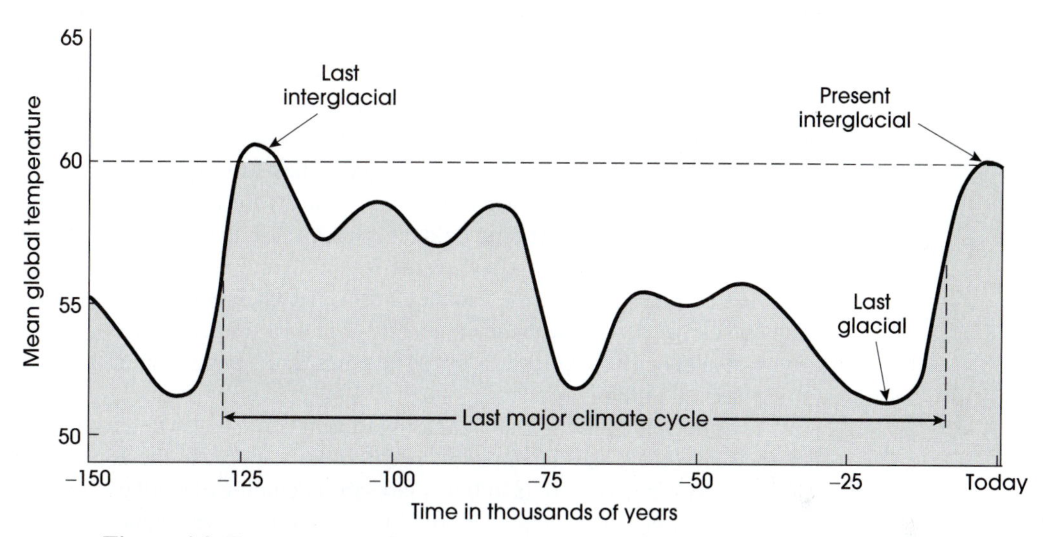


**Figure 4.2** Temperatures of the last 150,000 years. The last interglacial began about 130,000 years ago and ended about 110,000 years ago.

## *DATING THE PLEISTOCENE*

A problem concerning long-term climatic change such as those that took place during the Pleistocene, has been that of dating events. Geologists have long been able to provide relative dates, but absolute dating is much more difficult. Debate over the very age of the Pleistocene Epoch continued for over a century. In 1948, a group of scientists agreed that the beginning of the Pleistocene coincided with the appearance of certain cold-water species in rock found in southern Italy. Estimates by some geologists placed this event at 650,000 years ago.

Two developments aided the absolute dating process. One is the use of isotopes to provide dates. The other is the capability of obtaining samples of sediments from the seafloor. Generally, oceanic data are more reliable than continental data for dating events from the early Pleistocene.

The marine sediments are less disturbed than continental sediments. The oceanic data smooth the differences from place to place that occurred on the landmasses.

Plankton are microscopic organisms that live near the surface of the ocean where there is ample light. Each species lives in a certain temperature range. If the climate changes, they disappear and are replaced by plankton better adapted to the new conditions. When they die, their remains fall to the ocean bottom and become part of the sediment layers. These layers of sediment contain billions of microscopic skeletal remains of plankton. To get samples of undisturbed layers, cores up to 30 meters long are taken from the ocean floor. Analysis of the cores is a lengthy and tedious task. Twenty to fifty species and up to 500 individuals exist in a few centimeters of each core. Fossils found in the mud layers provide a record of temperature changes in the oceans, and in turn, in the atmosphere above the ocean.

Analysis of isotopes of carbon helps date the time an organism was deposited in the sediment. The microscopic skeletons contain both ordinary carbon and a minute trace of the isotope carbon-14. The proportion of carbon-14 to carbon-12 remains fixed while the organism is alive. After it dies, the carbon-14 begins to decay. By knowing the ratio of carbon-12 to carbon-14, one can determine the age of the shell.

Scientists reported on a group of cores from seafloor sediments that showed a clearly defined boundary based on changes in fossil remains. Using this sample, they placed the date of the beginning of the Pleistocene at 1.5 million years before the present. Although many geologists strongly protested this early date, it was soon pushed still further back in time.

Shortly after the turn of the twentieth century, Bernard Brunhes discovered that at some time in the past, Earth's magnetic field reversed itself. Later research shows that there were many such reversals, the last taking place some 700,000 years ago. The period since this reversal took place has become known as the Brunhes Normal Event. Isotope analysis of deep-sea cores helped date two earlier periods when the magnetic field was similar to that of the present. One reversal occurred at about 1 million years ago. The other, at the beginning of the Olduvai Normal Event, took place 1.8 million years ago.

The beginning of the Olduvai Normal Event corresponds to the appearance of the cold-water species in the Calabrian strata of southern Italy. Because the appearance of the cold-water species in the Calabrian strata in Italy marks the beginning of the Pleistocene and these fossils appear at the onset of the Olduvai Normal Event, the Pleistocene epoch is dated as beginning 1.8 million years ago.

## *EVIDENCE OF GLACIATION*

There is a greater degree of detail and accuracy for Pleistocene events than for earlier ice ages. The data are more numerous, more reliable, and there is a much greater variety. The number of proxy data sources increases as we examine events closer to the present. Marine shorelines, fossil pollen, ice cores, fossil soils, ancient shorelines, mountain glaciers, tree rings, and lake sediments provide information that is not available for earlier geological time (Figure 4.3).

### Evidence from Present Ice Sheets

Swiss scientist Louis Agassiz was a pioneer in formulating the theory that moving masses of ice modify the Earth. Beginning his research in Switzerland, Agassiz noted that the valleys had a U-shape rather than the typical V-shape of river valleys. Earlier researchers noted that U-shaped valleys contained very small rivers for the size of the valley. Most researchers explained this

**Figure 4.3** Glaciers greatly modified the land as they moved. In some places, such as this island in Lake Erie, they gouged out great scratches in the rock.

phenomenon as being the result of the biblical flood so vividly recorded in the Old Testament. These people believed that the large valleys were carved by much larger streams that existed when the flood receded.

Agassiz was also impressed by the presence of large boulders set amid assorted finer sediments that had obviously been transported from an area far from their present location. These boulders, called *erratics*, aroused much curiosity. Their presence was attributed to the biblical flood.

Other geologists did not agree that the biblical flood was the catastrophic cause for the erratics. They maintained that the forces that caused the U-shaped valleys and the erratics were the same forces that operate today. They maintained that present processes provide the key to what happened in the past. On this basis they thought the large erratics scattered over many areas of the world resulted from moving ice. They concluded that the erratics came from icebergs that floated on an extensive sea formerly covering large areas of Europe. Using the same reasoning, they decided that the assorted finer materials, called drift, in which erratics occurred, came from the same sea.

At first, Agassiz was convinced that the iceberg theory explained the erratics. Other scientists, including Jean de Charpentier and Ignace Venetz, also studied the landscape of the Swiss Alps. Some decided that the glaciers they saw had once been much more extensive and were responsible for the valley shapes, the drift, the erratics, and the parallel grooves or striations found on hard rock surfaces (Figure 4.4).

The accumulation of evidence of this type prompted Louis Agassiz to formulate a comprehensive theory of extensive glaciation. He suggested that a great ice sheet once extended

**Figure 4.4** Crevasse near the edge of Athabasca Glacier, in Banff National Park, Alberta, Canada. The crevasses develop due to uneven flow in the ice.

from the north pole to the Mediterranean Sea. The piles of rock rubble, grooves, erratics, and drift seen in Switzerland resulted from the action of this ice sheet. Thus the idea of an ice age was born.

## Information from the Ice

In the ice caps of Greenland and Antarctica lies a wealth of climatic information. Ice sheets form layer by layer from snowfall each year. With time, the snow changes into ice, often filled with bubbles of trapped air. By drilling into the ice and examining the vertical column, much can be learned about the environment when the original snow was deposited.

Drilling through the ice has become a high priority in research in reconstructing climate. Until recently, the deepest cores were drilled by Russian and French scientists at the Vostok base in Antarctica. On the Greenland ice cap U.S. and European scientists drilled two cores. They were the Greenland Ice Sheet Project 2 (GISP 2) and the Greenland Ice Core Project (GRIP), respectively. GISP 2 was a core 3000 meters deep.

The ice-core data provide an excellent record of climate over time. Among the atoms of oxygen there are some whose atomic weights vary from the normal atomic weight for oxygen. These are isotopes of oxygen. Most common are the heavy isotope, oxygen-18, and the lighter isotope, oxygen-16. The amount of oxygen-18 in water vapor is dependent on temperature. The warmer the air is when the water vapor formed, the more oxygen-18 there is. The ratio of the two isotopes in the ice depends on the temperature when the snow formed and fell on the ice. Thus the ratio of the two isotopes makes it possible to derive a pattern of temperature over time. Each layer of ice in the core has its own ratio of oxygen isotopes. During cold periods there is little oxygen-18 in precipitation. In the Greenland ice cap there is more oxygen-18 at the bottom and at the top. This corresponds to a warm period followed by a long cold spell and then a warming. Pollen analysis confirms the sequence. There is more pollen near the top and bottom of the ice, and less in the middle.

## Pollen Analysis

Relic pollen grains, or spores, are another form of proxy data. Many plants produce pollen grains in great numbers. For example, a single green sorrel may produce 393 million pollen grains and a single plant of rye, 21 million grains. The pollen spreads over the area in which the plants grow. Most important, the outer wall of the pollen grain is one of the most durable organic substances known. Even when heated to high temperatures or treated with acid, it does not visibly change. This is important, for pollen has shape characteristics that allow identification of plant groups above the species level.

To infer changes in climate, you need a layered sequence of pollen. This often occurs in ancient lakes or peat bogs where sediments cover seasonal pollen deposits. Cores taken of the sediments in these old lakes and bogs provide a sequence of pollen types. A high proportion of spruce pollen in the lower core may change to oak pollen at higher levels. This shows that a vegetation change occurred over time. Spruce grows in cooler environments than oak does, so the change from spruce pollen to oak pollen shows a change from cool to warm climatic conditions. Even a crude classification of pollen type, such as from trees to grasses, provides a rough guide to changing climatic conditions. The change from pollen associated with the nontree climates of the cold tundra to tree climates shows a warming of climate. The change from nontree vegetation to tree species in other environments indicates a change to a wetter climate.

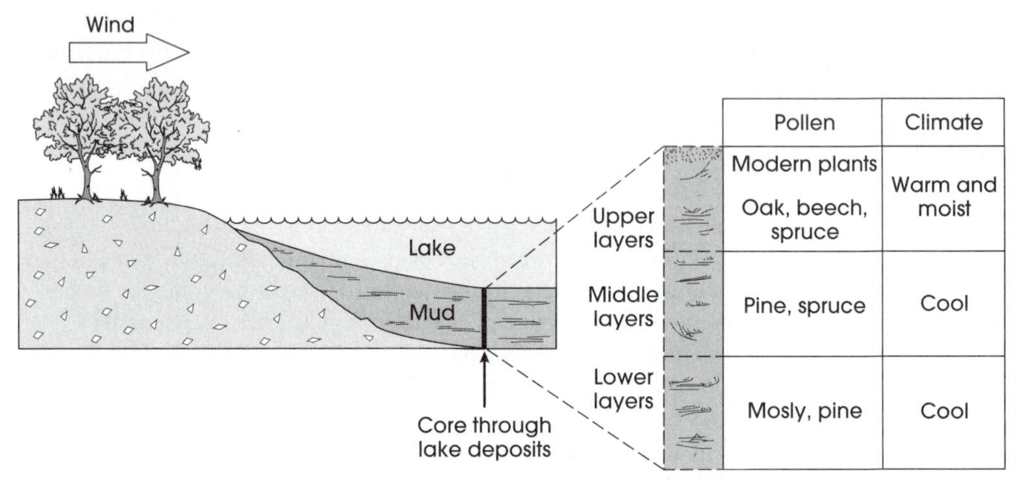


**Figure 4.5** Simplified diagram showing the method of reconstructing past climates using pollen analysis. (From J. E. Oliver, *Climatology: Selected Applications*, Silver Spring, Md. V. H. Winston and Sons. p. 223.)

Despite the important progress using this method, it does have shortcomings. A vegetation cover attains maturity only after a lengthy time, and it is feasible that the vegetation established through pollen analysis represents a successional stage that is not representative of the actual climate. In some areas there is a mixed vegetation cover and it becomes difficult to identify a dominant type associated with a particular climate. Also, plant communities do not always change their range as fast as climate changes. As a result, there may be a long lag in climate, as indicated by pollen analysis, compared to actual climate change (Figure 4.5). From Neolithic times, people have interfered extensively with the forest cover. Thus human-induced changes in vegetation might give misleading results.

### Other Evidence of Glaciation

Areas not directly affected by ice sheets also experienced a climate different from that of today. Some areas that are now quite dry experienced much wetter climates as a result of different weather patterns. Many inland basins contained large lakes as a result of the greater precipitation. There were many such lakes in earlier times.

The western United States, particularly in the Great Basin area, shows fine examples of the extent of such lakes. Lakes Bonneville and Lahontan, two such lakes, were enormous. At its maximum, Lake Bonneville occupied some 50,000 kilometers, an area about the size of Lake Michigan. The present Bonneville salt flats and elevated terraces are evidence of the former level of the lake (Figure 4.6).

## *THE LAST GLACIAL ADVANCE AND RETREAT*

The last major glacial advance in North America was the Wisconsin. Glacial ice was near its maximum extent as recently as 18,000 years ago. Ocean temperatures were lower than present temperatures by an average of 2.3°C. The high-latitude oceans were even colder. Over unglaciated land, air temperatures were 4.9°C cooler than now. Areas near the margins of the ice sheets in the

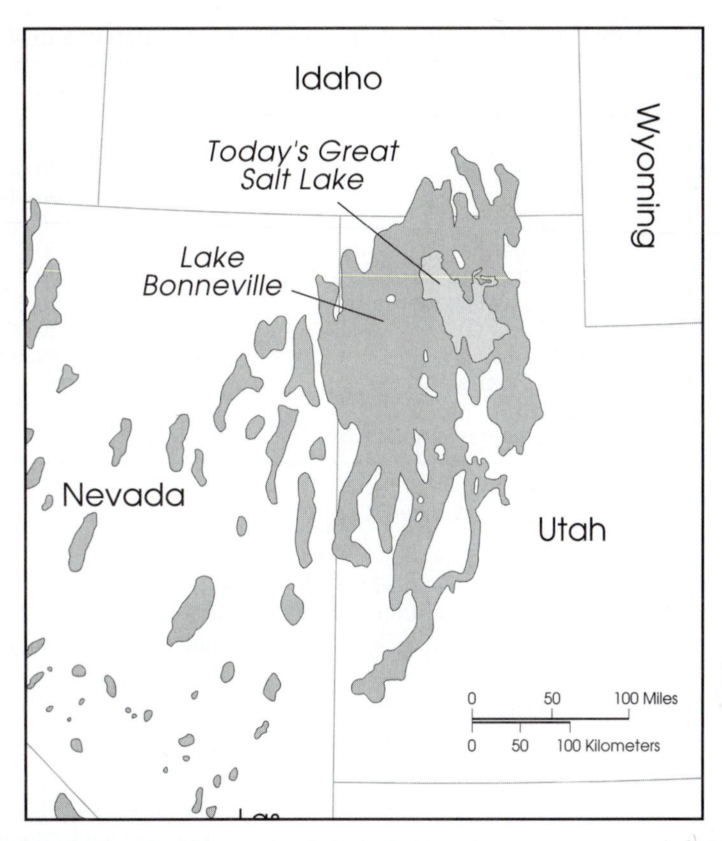


**Figure 4.6** During the Wisconsin glacial advance there were many rainfed lakes in the western basins. One of these was Lake Bonneville. Great Salt Lake is the remnant of this once much greater lake. The volume of water, and hence the area, of Great Salt Lake fluctuates over the years depending on the amount of precipitation in the surrounding mountains. The present size of Great Salt Lake is shown by the cross-hatching. Salt Lake City lies to the southeast of the present lake.

northern hemisphere showed the largest temperature differences. Here temperatures may have averaged 10° to 15°C less than now.

The Earth then was very different from what it is now. Much of the land was covered by ice. Ice covered 39 million square kilometers of Earth's land surface. This was 27 percent of the total land surface; only 3 percent is glaciated now. Ice covered Antarctica and Greenland as it does today and also covered large parts of the continents. Northern hemisphere ice probably reached a maximum extent of 32 million square kilometers compared to 2.3 million square kilometers today (Table 4.2).

In Europe there were ice centers over Scandinavia, Scotland, the Ural Mountains, and the Alps. At the maximum extent, ice covered nearly half the continent. The ice spreading from Scandinavia and Scotland joined to form a continuous ice sheet that covered all the islands except the very southern tip. The Scandinavian ice sheet spread southeast until it merged with another centered on the Ural Mountains of Russia. Another ice sheet formed in Siberia. The mountains of Asia supported glaciers that descended to low elevations.

**TABLE 4.2** Comparative Data for Present Glaciation and Maximum During the Pleistocene

| | *Present* | *Pleistocene* |
|---|---|---|
| Area ($km^2$) | | |
| Antarctica | 12,590,000 | 13,810,000 |
| Greenland | 1,800,000 | 2,300,000 |
| Eurasia | 170,000 | 500,000 |
| North America | 150,000 | 14,420,000 |
| North American Cordillera | 80,000 | 1,580,000 |
| Iceland and Spitzbergan | 70,000 | 440,000 |
| South America | 30,000 | 870,000 |
| Total area | 14,990,000 | 44,170,000 |
| Percent of Earth's surface | 3 | 9 |
| Volume ($km^3$) | $2.5 \times 10^7$ | $7.5 \times 10^7$ |
| Potential sea-level change (m) | −70 | −210 |

The southern hemisphere was not as different from what it is now. The maximum area of ice was about 13.3 million square kilometers compared to 12.7 million square kilometers at present. Glaciers existed in what are now tropical regions. Small mountain glaciers now exist on Mt. Kenya. During the peak of the ice age these glaciers descended 1600 meters lower on the mountain. The higher peaks of the Hawaiian Islands were also glacier covered.

In North America the centers of the ice sheets were the land areas east and west of Hudson Bay. From these centers the ice spread outward. It moved southward as far as what are now Long Island, the Ohio River, and the Missouri River. The ice sheet was thick. It was nearly 1 kilometer thick only a short distance from the edge. It was thick enough to bury the Appalachian Mountains in New England. The tops of many of these mountains show deep gouges left by the moving ice (Figure 4.7).

## CAUSES OF THE ICE AGES

It is not possible to explain with certainty the climatic changes of the Pleistocene. Some of the variables are known, but certainly not all. The most important factor in determining whether there will be a planetary growth of glaciers is the summer temperature in the high latitudes of the northern hemisphere. Here enough land exists to support continental ice sheets if the summers are cool enough. Winters are always cool enough to maintain ice and snow. What determines if there will be a continental ice sheet is how much of the winter accumulation of ice and snow melts in the summer. Therefore, prolonged periods of warm summers cause the ice to melt away. At latitude 65°N the warmer interglacial periods correspond to those times when summer insolation (solar radiation) was highest and the region presumably had warmer summers. Periods of cool summers cause the ice to advance as less snow melted away in the summer.

To explain repeated advance and retreat of ice sheets, there needs to be some mechanism to change the amount of energy received at high latitudes. One possible cause for the ice ages is a change in the output of energy from the sun. The energy emitted from the sun does vary through time. However, the magnitude of past variations would need to be about 10 times that of present variation to cause enough cooling for continental ice sheets to form. The changes in energy would need to last much longer than they do at present. There is no evidence to suggest that past glaciations correlate with changes in solar activity.

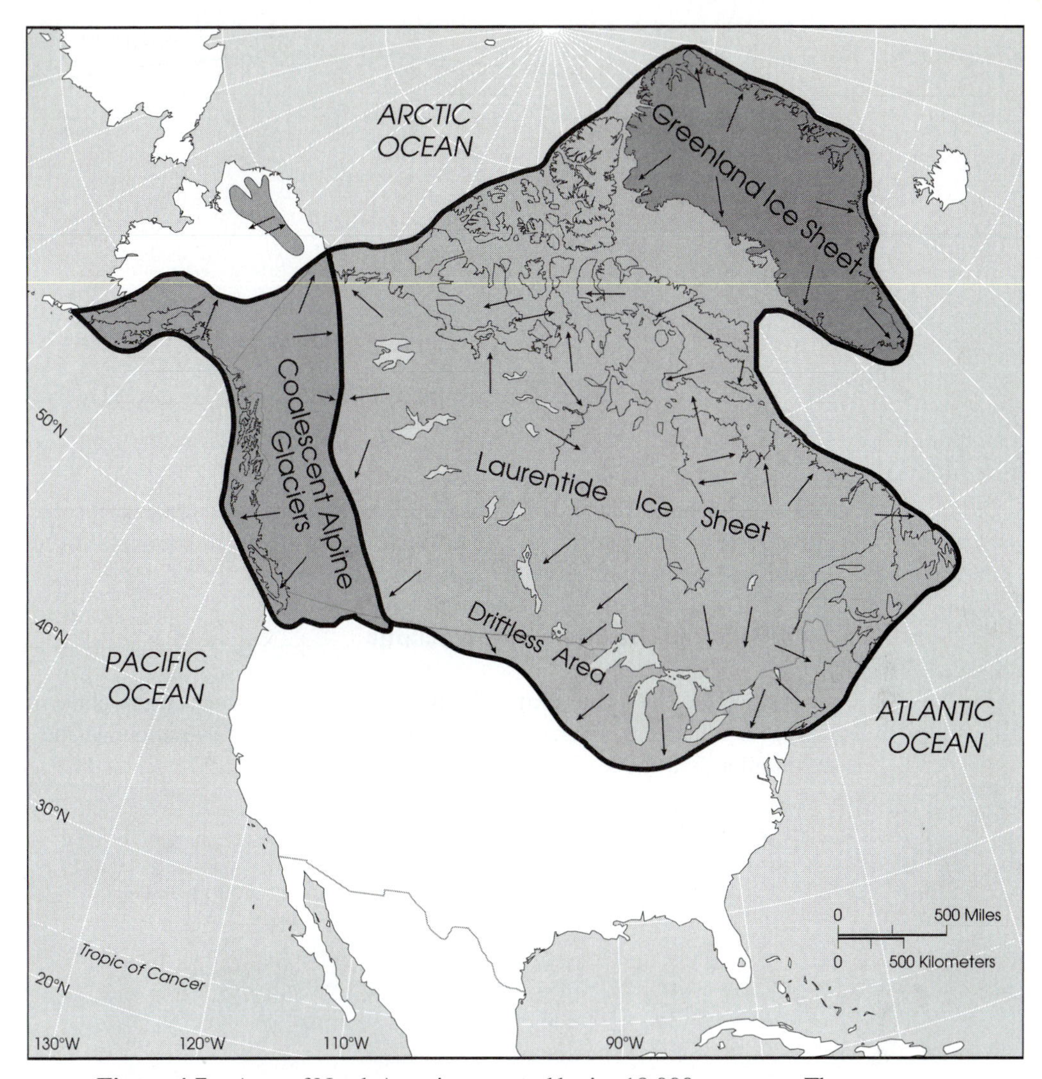


**Figure 4.7** Area of North America covered by ice 18,000 years ago. There were major centers of ice on either side of what is now Hudson Bay. The Greenland ice cap was more extensive than now. In western North America, mountain glaciers expanded to lower elevations and merged with the Laurentide ice sheet.

Related to glaciation are a series of movements of the Earth that produce regular changes in the amount of solar radiation reaching Earth and in the distribution of solar radiation received from pole to equator. One such mechanism involves a series of regular changes in the movements of Earth and sun. There are three aspects of Earth's position in the solar system that affect the distribution of energy at different times of year: the eccentricity of the Earth's orbit, the inclination of Earth on its axis, and the precession of the equinoxes (wobbling of Earth's axis).

## Earth's Orbital Eccentricity

Around the year 1860, James Croll suggested that the path Earth takes about the sun is not a circle but, instead, an elliptical path with the sun at one focus. The eccentricity is the difference in

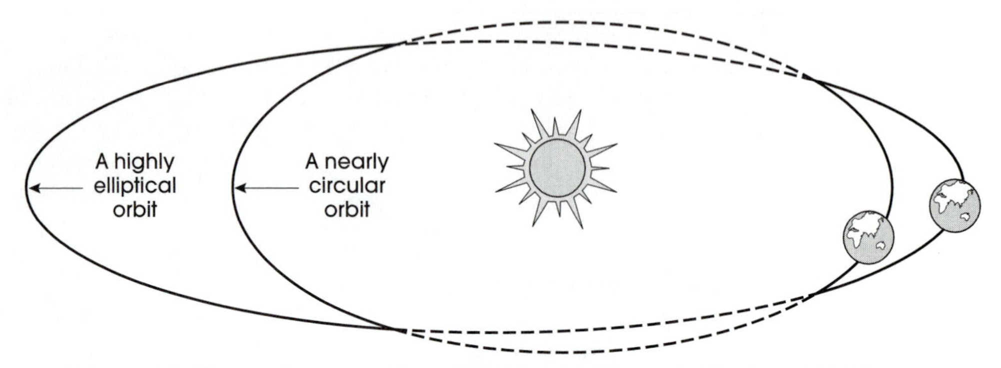


**Figure 4.8** Changes in eccentricity of Earth's orbit. The period is about 100,00 years.

the orbit from a true circle. The eccentricity influences the amount of solar radiation intercepted by Earth and changes the dates at which the solstices and equinoxes occur (Figure 4.8). These dates are the beginnings of our four seasons.

### Obliquity of the Plane of the Ecliptic

The angle of the plane formed by the geographic equator and the plane in which the Earth revolves around the sun changes through time. The angle varies around a mean value of 23.1°. The variation is from 21°59′ to 24°36′ over a period of 41,000 years. The present angle is about

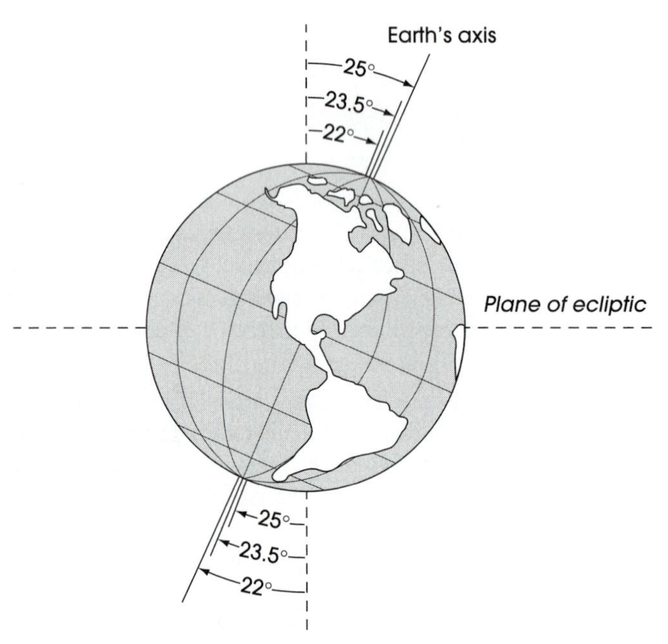


**Figure 4.9** Changes in the angle of Earth's axis with the plane of the ecliptic. The period is 41,000 years.

23°27′. In 1904, astronomers hypothesized that the change in obliquity has far-reaching effects on climate. The greater the angle between the two planes, the greater the contrast between summer and winter in middle and high latitudes. An obliquity of 0° results in equal lengths of day and night over the globe all year and a lack of seasonal changes throughout the year. A high angle results in extremes in the lengths of summer and winter days and nights. Although the actual changes in the angle of obliquity are not large, they are enough to cause distinctive changes in the latitudinal zonation of climates (Figure 4.9).

## Precession of the Equinoxes

The time of the year when Earth reaches its closest and farthest distance to the sun varies. This is due to the position of Earth with respect to the other planets and the nature of its path around the sun. This shift is important in the radiation balance of Earth. Most of the land area is in the northern hemisphere, and land and water respond differently to incoming radiation.

J. F. Adhimar published a book in 1842 entitled *Revolutions of the Sea*. In his book he theorized that the ice ages were a result of the precession, which has a period of 23,000 years (Figure 4.10). When Earth is closest to the sun in the northern hemisphere summer, it increases the difference between the seasons. In 9300 B.C. (11,300 before the present), perihelion was on June 21 and the northern hemisphere experienced warmer summers and cooler winters. When Earth is closest to the sun in the northern hemisphere winter, it reduces the adverse conditions of winter and summer. In A.D. 1200, perihelion occurred on December 21, so the southern hemisphere received its maximum radiation in summer.

Now perihelion occurs during the northern hemisphere winter, but 10,500 years from now the date of perihelion will occur in the northern hemisphere summer. Then the lowest input of so-

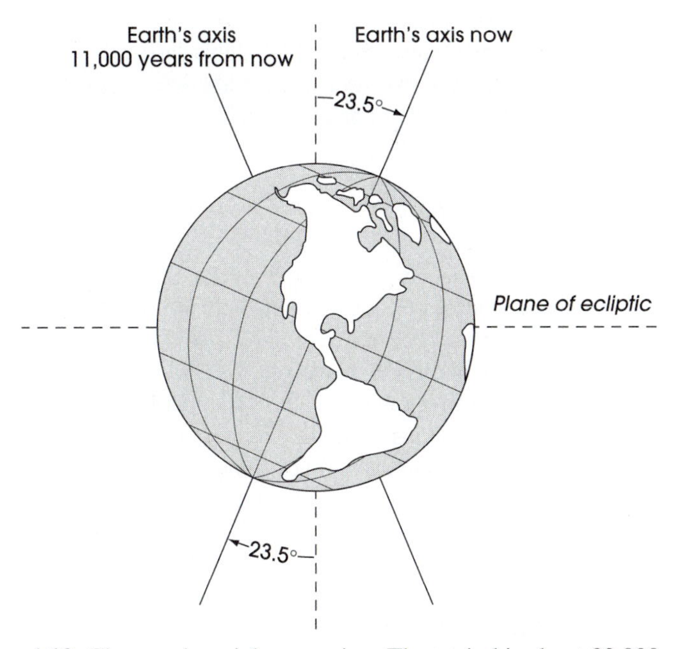


**Figure 4.10** Changes in axial precession. The period is about 23,000 years.

lar radiation will occur in winter in the northern hemisphere, where most of the land is found. Since land cools more than sea, extremely cold temperatures will occur.

Precession, obliquity, and eccentricity are periodic through time. The state of each can be calculated for any time in the past or future. It is possible to determine the times when the various cycles of change reinforced one another to produce very high or very low radiation values in the northern hemisphere. Long before high-speed computers, the Yugoslavian scientist Milutin Milankovitch derived values going back thousands of years.

Many climatologists think that these orbital variations are the basis of climatic change. The orbital parameters were such that summer radiation was last at a maximum 128,000 years ago. Data showing changes in coral reefs place the last interglacial at 122,000 to 130,000 years ago. This matches well with what the orbital parameters predict for that time. The extensive glaciation during the last glacial advance corresponds to a time when the Malankovitch cycles produced a minimum of seasonal differences in high latitudes in the northern hemisphere and hence reduced summer radiation at 65°N. Gradually, the system changed to produce maximum seasonal variation once again. The glaciers retreated in the warmer summers.

### Orbital Parameters as Insufficient Cause

It is evident that the orbital variables are not sufficient to explain the ice ages. There must be other driving forces. During significant periods of Earth history no major glaciations took place when the changing orbital parameters show that ice sheets should have existed. For example, only two major ice ages developed in the last 600 million years. The orbital conditions suggest that there should have been many more. Glacial advances and retreats lag maximum and minimum periods of radiation. Continental ice sheets may continue to melt well after conditions have occurred that will eventually result in expansion of the ice sheets.

The Milankovitch cycles explain only about $\frac{1}{2}$° out of a mean difference of 4°C in global temperature between glacials and interglacials. Over the 100,000-year period, energy received on the planet changes only $\frac{1}{2}$ percent due to eccentricity. The shorter cycles, on the other hand, change the amount of energy reaching the higher latitudes by as much as 20 percent. These cycles are a part of the mechanism producing the alteration between glacials and interglacials, but there must be other processes that operate as well.

## *OTHER CAUSES OF GLACIATION*

One element that may play a part in glaciation is the latitudinal distribution of the continents. When Pangea existed there was essentially only the one landmass. This would have permitted much freer flow of water in the oceans. The unrestricted flow of water would carry large amounts of heat into polar regions. For this to take place, there would not need to be any change in the global supply of energy. The problem with this theory is that Pangea was glaciated extensively during the Mesozoic. The land making up India, southern Africa, and South America was extensively glaciated.

Changes in atmospheric composition may also play a part in glaciation. Modern volcanic eruptions have placed large quantities of dust and sulfer dioxide ($SO_2$) into the atmosphere. In theory—in some cases borne out by volcanic eruptions—the dust cloud that results blankets the Earth and reduces sunlight. However, the dust veils have lasted for a few years at most and have not been sufficient to reduce global temperatures far enough or long enough to cause a change in the ice mass.

When changes take place in seasonal temperatures from year to year, rates of evaporation and precipitation change as well. This, in turn, may cause ocean currents to change. When ocean currents change location, heat is distributed differently over the Earth. Sudden and sharp changes in the major ocean currents amplify changes from glaciation to interglacial periods. When the currents carry heat poleward, the ice sheets melt. When the currrents cease to flow poleward, the ice sheets grow.

## *SUMMARY*

The environment of the Pleistocene consisted of glacial and interglacial periods. Ice sheets advanced into warmer regions and then retreated. The most recent full glacial advance, known as the Late Wisconsin in North America, lasted from perhaps 30,000 to 12,000 years before the present. The coldest temperature was 4 to 6°C (7 to 11°F) lower than present and occurred about 18,000 years ago. Ice sheets extended as far south as 50°N in Scandinavia and 36°N in North America. Frigid polar water extended in the North Atlantic to 45°N. Earth's environment at this time was very different from that of today.

The causes of the ice ages are uncertain. It is clear that many environmental factors play a part. These include changes in the amount and distribution of sunlight over the planet, atmospheric composition, distribution of the continents, and ocean currents. Other elements may also be involved.

## BIBLIOGRAPHY

BROECKER, W. S., and G. H. DENTON. 1990. What drives glacial cycles? *Scientific American*, 267:49–56.

DAWSON, A. G. 1991. *Ice Age Earth.* New York: Routledge.

WILLIAMS, M. A. J., P. DE DECKKER, A. P. KERSHAW, and J. A. PETERSEN. 1993. *Quaternary Environments.* London: Edward Arnold.

# chapter 5

# Early Growth and Spread of the Human Population

***CHAPTER SUMMARY***



It is uncertain exactly how long the human species has been present on earth. The date of appearance is being pushed further and further back in time. The earliest forerunners of the species go back beyond the Pleistocene Epoch into the Tertiary. In the earliest stages of human evolution, several physiological and structural changes took place. A prehensile hand and a flexible arm developed together with stereoscopic vision and color perception. Later, an upright posture appeared, which freed the hands for other tasks. It is bipedal locomotion that is now used to distinguish between hominoids and other animals. The divergence of the hominoids from the gorilla and chimpanzee is most likely to have taken place around 5 million years ago.

In 1974 anthropologists found fossilized teeth and jaws from 12 hominoid individuals, and in 1976, discovered fossilized footprints in volcanic ash in the Laetoli area of Tanzania. The footprints were left by two separate individuals at different times. The volcanic ash eventually hardened into rock and preserved the footprints. They were clear enough to tell that one of the individuals stopped, turned to the left, and then went on his or her way. The relics are 3.6 to 3.8 million years in age. Estimates made from the size of the foot and length of stride gave a height of 123 centimeters for one of the individuals and 143 centimeters for the other. The prints were similar in shape and function to humans of today, with rounded heel, raised arch, and a large big toe not found in other primates.

Also in 1974, Donald Johanson and Tim White found the fossilized remains of a female skeleton that was dated between 3 and 4 million years before the present. Lucy, as she was nicknamed, stood about 107 centimeters and weighed about 23 kilograms. She walked upright and was proportioned similarly to modern people. A year later the same group, working in the Afar Triangle of Ethiopia, found a stratum containing the fossilized remains of 13 individuals who apparently perished at the same time. This group, obviously in a very primitive stage, left no evidence of tools. Other evidence gathered in East Africa corroborates this early appearance of the human species.

## *HOMINOIDS AS FOREST DWELLERS*

In the process of evolution, hominoids did not do well as forest creatures. As the limbs necessary for efficient climbing, such as a prehensile tail, were lacking, they were of limited effectiveness in the forests. Chimpanzees and other members of the ape family, whose diets are primarily fruit, are able to digest unripened fruit. This is not true of the human species. When food was scarce in the forest, the species was forced to seek food on the ground. The species is a frail competitor among carnivores and is extremely poor compared with such specialized hunters as the cheetah or wolf. The species lacked, and still lacks, claws, fangs, or enough speed to be an effective carnivore. Also missing is the dentition and digestive equipment required to be an effective herbivore. Hominoids were essentially scavengers of meat and gatherers of plants. Much sustenance was probably from highly concentrated nutrient supplies such as fruit, seeds, bulbs, and tubers. In all likelihood, the species did best in open woodlands, where a great variety of plants and animals were found. Perhaps the tropical forest edge was the place of origin, due to favorable temperature, moisture, and the variety of foods available. The high demand for fresh water limited activity in the desert regions, although small numbers may have existed there as they do at the present time. At this time, humans were perhaps intermediate in the food chain, being preyed upon by more powerful carnivores and preying on lesser creatures.

The size of territory the group needed depended on limiting environmental factors, of which rainfall was one, particularly in the tropics. In areas where rainfall was less than 60 centimeters a year, territorial demands were very large. The demands for space decreased rapidly as rainfall increased above this level. Hominoids had little control over their environment. They must have been a daytime creature that sought shelter at night for protection from cold and predators. The main source of sustenance was wild food over which there was no control. The changing seasons must have meant seasons of plenty and seasons of scarcity. The availability of food was a major constraint on the size of the population. When food supplies were abundant, the population expanded. When food supplies became scarce, there was a dieback of population. This form of feedback mechanism still controls many animal populations today.

During this long, early period, the species was surrounded by the same material environment as at present, however most of the resources we have at present were not available to Paleolithic man as the knowledge of how to use them did not exist. Males must have spent their whole time hunting and scavengering. Women were no doubt fully occupied in food collecting and protecting their young. The mortality rate must have been extremely high and the expectation of life very low. Probably many of the most enterprising men, the daring hunters were killed off at an early age. The elderly and infirm would certainly not have been supported, and those that could not keep up with the rest would have been abandoned.

## *DEVELOPMENT OF TOOLS*

The human species has always had an impact on the environment. Individuals and groups have taken from the environment what they needed and what they could get to sustain life. Technology has always played a part in the extent of the impact. Even simple technology can make a notable difference in the extent of change an individual or group can make.

The first great leap for the species came with the development of tools. The oldest stone tools used for cutting and scraping date from 2.5 million years before the present. With the development of stone weapons, the standing of the species rose to that of a middle-sized predator near the top of the food chain. The early hominoids were most successful when hunting in small packs. Even at this early time the species began to create and destroy resources. It is inherent in resource creation that there will be some form of resource destruction, if only in the form of direct consumption.

Among the earliest of created resources were the chert and flint supplies used to make weapons and tools. Some of the earliest settlements were probably based on the control of supplies of these useful rocks. The use of sharpened rocks for hunting implements increased hunting skills, allowing hunters to take more of some animal species and to add more wildlife species to their food supplies. There is no evidence to support the notion that these early peoples held any different attitude toward the environment from that of the present. There is ample evidence that groups often upset the balance of their environment by overkilling and were forced to move or died of starvation. Thus the creation of a resource in the form of better weapons led to the destruction of other resources: namely, food supplies.

## AUSTRALOPITHECUS BOISEI *AND* HOMO HABLIS

Early in the Pleistocene, at least two main hominoid groups were present, the older and more primitive Australopithecinae, and a more advanced form, *Homo hablis*. *Australopithecus boisei* had an essentially humanlike body, limbs, hands, and feet. The head, however, was still apelike. They were not makers of tools, but used sticks, stones, or bones, as they were found, for crude tools.

Simple technology has been accepted as the threshold of the human species. *Homo hablis*, the more advanced species, was a user of pebble tools, which they made in a uniform fashion. Their tools exist in association with animal bones that show signs of deliberate cutting and breaking. The species was relatively short-lived, existing in Africa about 2 to 2.5 million years ago. *Homo hablis* is also the first hominoid that has a measurably larger brain case than that of other primates. Facial features, jaws, and teeth were similar to those of the earlier *Australopithicus*.

## HOMO ERECTUS

A species that overlapped *Homo hablis* and perhaps *Australopithecus*, but developed much further, was *Homo erectus*. *Homo erectus* was intermediate between the apes and humans. The first remains were found in 1890 near Trinil, Java, in the bank of the Solo River. The find consisted of a skull cap, a few teeth, and a thigh bone. Nearly 50 years later a skull and other pieces of bone were found that belonged to several different individuals of the species. In 1974, Richard Leaky, Director of the National Museum of Kenya, found the skull and other bones of a young male who lived 1.6 million years ago. This was the most complete human skeleton yet found dating from such an early period, and it was of the species *Homo erectus*. Study showed that the individual died at the age of 12 years and was about 164 centimeters tall. In all likelihood, had he reached adulthood he would have reached nearly 169 centimeters in height. The find included a skull, collarbone, vertebrae, ribs, limbs, and pelvis. Some 70 pieces were found over a 6-meter-long space between beds of volcanic ash. The age was determined by the structure of the teeth, the height by the length of the thigh bone, the sex by the shape of the pelvis, and the species by the shape of the skull.

The head was characterized by a low brain case, flat forehead, heavy protruding eyebrows, a large and thrusting upper jaw, a heavy lower jaw without a significant chin, large teeth, slightly longer canine teeth, and a cranial capacity of 700 to 800 cubic centimeters.

Further evidence of *Homo erectus* was found in Kenya in 1975 with the location of a skull that dated from 1.5 million years ago. *Homo erectus* represents a distinct stage in the evolution of modern humans and may have persisted until as late as 400,000 years ago.

## *FIRE AS A TOOL*

Perhaps the second big shift in the balance between the population of humans and the environment came with the control of fire. By 1.7 million years before the present, Yuanmo Man lived in what is now Yunan Province in southwestern China. These people used both stone tools and fire. The use of fire opened up new resources in the form of new tools and new food supplies. Fire was used both as a source of heat for warmth and for cooking. Fires were set by hunters to flush game from undergrowth and to eliminate undergrowth too thick to allow hunting. The human species was still limited to wild food, but it emerged as hunter of most of the animal kingdom that was deemed edible. This provided a much greater variety and amount of food. As a result of the increase in food supply, the population took a sharp upward turn. It probably reached a total of 125,000 by 1 million years before the present. As slow as progress was during this period, significant changes did take place.

## HOMO SAPIEN NEANDERTHALIS

*Homo sapiens neanderthalis* developed during the third interglacial stage and survived until well into the last glacial stage. Remains of more than 300 individuals have been found scattered through Europe, Asia, and Africa. Most of the finds have been in Europe. The earliest remains date from about 300,000 years ago. Most date from 100,000 to 45,000 years ago. These people had very sturdy bones, large chests, large feet, receding chins, and heavy brow ridges. Their cranial capacity ranged from 1100 to 1155 cubic centimeters.

The Neanderthals were largely nomadic, following their food supply. It seems equally certain that some were sedentary where food, water, and shelter were available year round. All of these were critical items. They may have begun to fabricate shelters from hides. They were crafting shoes, caps, and capes, so a simple tent should have been no problem. Neanderthal people became expert workers of stone, being the first to attach flint points to spears. They also manufactured hand axes, spear points, scrapers, cutting blades, and awls.

Many Neanderthal fossils have been found associated with glacial deposits, indicating that these people lived in close proximity to the ice sheets. Near the edge of the glaciers, and especially as the glaciers retreated, grasses and shrubs would have prevailed for many years before

**Figure 5.1** The continental ice sheets had abrupt fronts that advanced and retreated. The glacial meltwater formed huge rivers. The Columbia River Valley, Lower Mississippi River Valley, and the Ohio River Valley are examples of valleys that were carved by much greater rivers.

the forests reached maturity. Grazing animals, large and small, depend on grasses and shrubs for food. Thus, with an ample supply of food and water, the areas around the perimeter of the ice sheets must have sustained large numbers of grazing animals. It is well established that these people were skilled hunters of game of all sizes, including horses, woolly rhinoceroses, and mammoths. The animals they killed provided skins, sinews, horns, and bones.

## HOMO SAPIEN SAPIEN

Molecular biology and analysis of DNA have sharpened the debate and at the same time clarified the history of the human species. Modern humans or *Homo sapien sapien* evolved in Africa some

**Figure 5.2** A boar hunt with spears and dogs depicts how the Cro-Magnon of 12,000 years ago had turned from larger herd animals to small game. Source: The Field Museum, Chicago. #A111668

200,000 years ago (Figure 5.2). From here they spread out over the planet, displacing other hominoid forms. It appears that *Homo sapien sapien* replaced the Neanderthals. Neanderthals became extinct first along the southeastern fringe of their range 45,000 years ago and disappeared from the fossil record in Europe some 32,000 years ago. Ezra Zubrow of the State University of New York believes that only a small competitive advantage was enough for modern humans to be able to displace the Neanderthals. He believed that replacement could have occurred in as little as 30 generations. Just by occupying the best dwelling sites and having a slight edge in hunting technology the modern humans could have forced the Neanderthals out.

*Homo sapien*, or Cro-Magnon Man, represents a more advanced hunter and gatherer than the Neanderthals. The name *Cro-Magnon* comes from the area in France where remains were first discovered in 1868. Cro-Magnon people predominated from 40,000 years ago until the advent of agriculture 10,000 to 11,000 years ago. These were relatively large people, some being at least 184 centimeters tall, with a high forehead, small brow ridges, narrow prominent noses, and highly developed chins. The cranial capacity reached 1700 to 1800 cubic centimeters. While the largest numbers may have lived in Europe, they had spread over most of Africa and Eurasia by the time the great ice sheets began to melt.

They hunted mammoths, bison, wild cattle, rhinoceros, horses, and deer. These were all found on the margins of the great ice sheets. A large camp used by mammoth hunters was formed along the Pechora River in the former Soviet Union. Nearly all of the bones were those of mammoths. The remains of a structure made of mammoth bones was also located. These people used highly developed stone tools as well as tools made from bone. Evidence of clothing dating from 30,000 years ago was found. The garments consisted of trousers with boots, a pullover shirt, and head covering. The items had also been decorated with ivory beads.

## *HUMAN OCCUPANCY OF NORTH AMERICA*

In North America the history of primates and humans is clearly different from that of the other continents. Primates that had existed on the continent disappeared by the middle of the Tertiary Period. No fossils of primates have been found that date from the beginning of the Miocene until almost the end of the Pleistocene. From the beginning of the Miocene the western hemisphere was isolated from the eastern hemisphere. The isolation continued until the ice sheets of the Pleistocene lowered sea level such that a land bridge appeared in the Bering Strait. Predecessors of the Eskimos may have crossed the land bridge at the Bering Strait as early as 100,000 years ago, establishing themselves north of the great ice sheets in Alaska and western Canada. The physical characteristics of the Eskimos suggest an origin in Asia. The earliest arrivals brought with them fire- and toolmaking skills.

The earliest indications of human occupancy of northeastern Asia appear about 25,000 years ago. About 50 meters below present sea level is a shelf across the Bering Strait. The migration route from Asia to North America lay across this shelf when it was exposed by a lowering of sea level. The shelf was exposed several times during the Pleistocene (Figure 5.3). Early in the Pleistocene many land animals crossed this bridge. Among them were the mammoth, mastodon, bison, saber-toothed tiger, short-faced bear, dire wolves, and cheetah. Sometime during the last glacial advance this land bridge was exposed. It may have been exposed at intervals and perhaps a good deal of the time. About 18,000 years ago sea level was at least 100 meters below that at present.

Once across the land bridge the route south was still blocked by ice. The Cordilleran ice sheet over the Rocky Mountains and the Laurentide ice sheet were fused together east of the Rocky Mountains. Both of these ice sheets were immense in area and depth. The combined ice sheet reached from the Pacific Ocean to the Atlantic Ocean and reached 2500 meters in depth over large areas. At the time of the last glacial maximum 18,000 years ago, the route south from the arctic shore was surely blocked.

Despite their early appearance along the arctic sea, humans did not spread across the Americas in numbers until after the last ice sheets began to melt away. As the Rocky Mountain glaciers and the great continental ice sheets began to retreat, a dry corridor opened up through the middle of the continent perhaps as early as 15,000 years ago.

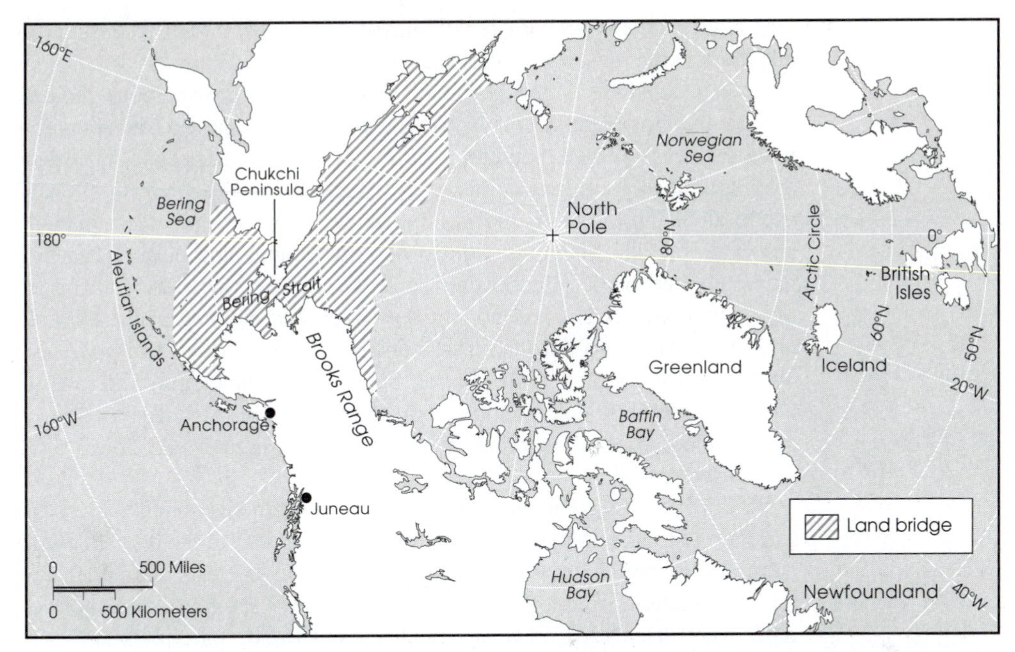


**Figure 5.3** During the height of the Wisconsin glacial advance, sea level was much lower than at present. A large portion of the continental shelf between Asia and North America was exposed above sea level. This is shown as the crosshatched area on the map. The land bridge between the two continents allowed the migration of people and animals back and forth.

Many archeological sites in the Americas have been dated from 10,000 to 12,000 years ago. In a cave in southeastern Washington state were found the remains of human bones, clothing, remnants of fish and shellfish, and the charred bones of members of the deer family. The remains, referred to as Marmes Man, date from 13,000 to 14,000 years ago. There are numerous sites found throughout the Great Plains and western mountains which date from 11,500 to 11,000 years ago. These sites contain fluted projectile points characteristic of the Clovis culture. A site in Mexico containing a human skeleton, stone artifacts, and bones of extinct mammoths is in excess of 10,000 years old. In South America, sites have been found that were occupied at least 12,000 years ago. It seems clear that once people moved onto the North American continent in substantial numbers, they spread rapidly southward.

## THE PLEISTOCENE EXTINCTION

In the recent history of the Earth there have been two periods of accelerated extinction of biota during which the extinction rates were greater than evolutionary causes would predict. These two periods are the end of the Pleistocene Epoch and the modern period. The Pleistocene extinction took place over a period of a few thousand years at the end of the last glacial advance (18,000 to 8000 before the present).

In a relatively short period of time, whether measured in geologic time, evolutionary time, or in terms of human history on Earth, dozens of species of mammals and birds disappeared without being replaced. Something unusual must have happened. The spacing of the extinctions over several thousand years makes it difficult to ascribe them to climatological or geological events alone. Furthermore, most of the extinct animals were large—mammals over 50 kilograms and birds certainly large enough to be conspicuous and edible. The largest North American mammals were hardest hit. Seventy percent became extinct during the last 10,000 to 15,000 years of the Pleistocene. Horses and camels, which had evolved in the New World, became extinct on this continent, as did mammoths and mastodons, which had migrated into the New World over the once dry land of the Bering Strait. The ground sloth, the saber-toothed tiger, the dire wolf, the giant buffalo, antelope, and the giant beaver also disappeared, yet there was no concomitant loss of small mammals, plants, or aquatic organisms. Table 5.1 dates the disappearance of selected species.

It may very well be that humans became proficient enough hunters to be instrumental in the extinction of large animals that took place. The spread of the human species and the state of the cultural development at the end of the Pleistocene suggests that humans were a major factor in this accelerated extinction. A surprising number of the fossilized bones of extinct animals and birds were discovered associated with charcoal and stone tools such as arrowheads or spear points. Some of the sharpened tools were still embedded in the bones.

It was long believed that large mammals were too powerful for humans to hunt. Perhaps this was true for the individual hunter with spear or bow and arrow, but when hunters went out in groups, it is quite possible that they could have been very successful hunters of large animals, even using only simple tools. The Plains Indians of North America demonstrated that they could hunt large animals successfully with primitive weapons. Herds of animals were driven off cliffs or into water by the use of fire. They were either destroyed in the fall or drowned, or at least made far more vulnerable to attack. In Africa, pygmies hunted elephants with primitive tools. They covered themselves with elephant dung to cover up their scent and make it easier to approach their prey. When they came within striking distance, they thrust their spear into the soft underparts of the animal and then withdrew and followed the animal until it collapsed. To speed its death, they would cut off the trunk, causing it to bleed to death.

It is also likely that early humans hunted large animals by surrounding the animals with a ring of fire. In Africa, until quite recently, elephants were hunted by surrounding them with grass fires. When they attempted to break out they were driven back by shouting or by beating on drums. Animals soon succumbed to the smoke and heat or were weakened and terrified to the point where they could be killed more easily. One of the simplest cooperative techniques was that practiced in Tasmania. The hunters made a huge circle, so big that the individual members of it were only just within shouting distance of each other, and at a given signal moved forward, shouting and making as much noise as possible. By this means encircled game were frightened and gradually driven into a more and more confined space. Eventually, the ring of hunters became so constricted that they were able to link hands. Then the spears were thrown and the clubs wielded until the scene became a chaos of blood, screams, and bellows as the confused, panic-stricken animals were butchered.

The South African bushpeople also used this type of hunting, but in a more elaborate form that required careful preparation. A valley was selected and across it a fence was built. Gaps were

**TABLE 5.1** Extinction Dates of North American Mammals

| *Years before the present* | *Species* |
|---|---|
| 6,000 | Mastodon |
| 7,800 | Colombian mammoth |
| 8,000 | North American horse |
| 9,500 | Ground sloth |
| 10,500 | Woolly mammoth |
| 14,000 | Saber-toothed cat |

left here and there, but with large pitfalls dug out of them and with hunters concealed nearby. The remainder of the tribe moved out in a wide circle to beat the game in toward the fences. On reaching it, the animals would run along until they found a gap, but when they tried to pass through, they fell into the pits and could be slaughtered at the hunters' leisure.

It seems quite possible that in a time of severe environmental stress caused by the glaciation, certain animal populations may have been pushed beyond their point of survival by human hunting activities. If their numbers had already been reduced by unusual natural stresses, these group hunting practices could have led to their extinction. Humans appeared in North America rather suddenly from Asia, occupied the continent, and began hunting animals that had never had a chance to adapt to this special type of predator. In Africa, far fewer of the fauna were lost during the same time in history. This lower rate of extinction may have been due to the concurrent evolution of humans in Africa, giving some of the animals time to develop defenses against human predation.

## SUMMARY

During the Paleolithic period the human species was struggling against the environment. The size of the population was limited by the amount of wild food that could be gathered. When food was scarce the population died back, and when it was plentiful the population expanded. Diseases resulting from technology began even at this early date. Sinusitis is a disease that irritates the nose as a result of breathing damp, smoky, or dusty air such as would be found in caves. The disease has been detected in skulls dating back well into the Paleolithic.

Only limited numbers of human lives could be lost from storms and other short-lived events, as the population was small and widely scattered. The population growth rate during the first 500,000 years was very slow. Birthrates were very high, perhaps between 38 and 42 per 1000. The death rate was also very high, perhaps 35 to 38 per 1000. Infant mortality was high and the average life span was only about 30 years.

Although life was difficult for hominoids during this time, the human species triumphed. Charles Darwin stated the case very well: "Man in the rudest state in which he now exists is the most dominant animal that has appeared on this earth. He has spread more widely than any other highly organized form; and all others have yielded before him. He manifesly owes this immense superiority to his intellectual faculties, to his social habits . . . and to his corporeal structure. The supreme importance of these characteristics has been preved by the final arbitrament of the battle for life" (Darwin, 1871).

## BIBLIOGRAPHY

DARWIN, C. 1871. *The Descent of Man and Selection in Relation to Sex*. London: John Murray.

chapter 6

# The Holocene: Earth's Environment Since Deglaciation

***CHAPTER SUMMARY***



The Holocene is the last, and current, geological epoch. It is also the shortest. Holocene means "entirely recent." By 11,500 years ago, only scattered areas of ice sheets remained in western North America. The main ice sheet was in eastern Canada. Real warming of the northern hemisphere could not take place until most of the ice melted. This is because of the large amount of heat it takes to change ice to water. It takes 80 calories of heat to melt 1 gram of ice without changing its temperatures. This is as much heat as it takes to raise the temperature of a gram of water from freezing to 80°C. Several interconnected events mark the boundary between the Pleistocene and the Holocene.

## *THE YOUNGER DRYAS EVENT*

Toward the end of the long retreat of the ice, the climate flipped back and forth between warm and cold. The ice sheets began to melt rapidly about 14,000 years ago. During the initial phase of melting, the meltwater in North America flowed southward into the Missouri and Ohio rivers. The

Mississippi River then carried it into the Gulf of Mexico. The volume of meltwater must have been nearly as large as the present flow of all the rivers on the continent. The fresh water affected both the salinity and temperature of the Gulf of Mexico. The cold water spread out in the Gulf of Mexico and warmed from the heat from the sun.

As the sea ice melted in the Atlantic Ocean, the edge retreated northward. Warm water from equatorial regions moved farther north in the Atlantic Ocean. The clockwise circulation around the North Atlantic, including the gulf stream, established itself, setting up weather patterns similar to those of today.

About 11,500 years ago, a strange event occurred. The margins of the remaining ice sheet expanded and some small ice sheets reappeared. This has become known as the Younger Dryas Event (named for a small flower found in cold climates). This was not a global event. It took place in and around the North Atlantic Ocean. Evidence of the cold is in seafloor sediments and in glacial ice on Greenland, maritime Canada, and northern Europe. It was the last major swing to cold weather around the North Atlantic Ocean.

Centuries later, the North American ice sheet retreated north of the Great Lakes. An outlet appeared for meltwater to flow directly into the North Atlantic instead of into the Gulf of Mexico. This opening is now the present Saint Lawrence River system. The large flow of cold water spreading out over the North Atlantic stopped the flow of the warm gulf stream. The gulf stream was no longer carrying massive quantities of heat into the area. The result was a sudden and drastic drop in the temperatures around the North Atlantic. Arctic pack ice spread rapidly to the south, further chilling the atmosphere above. During this cool period forests died out, especially those in Europe. Tundra vegetation, more typical of colder climates, replaced the forests.

## *WARMING SETS IN AGAIN*

The return to the ice age climate ended abruptly around 10,700 years ago, having lasted for eight centuries. The change from the cold to a warmer and milder climate took place rapidly. This is especially so if measured in geologic time. The volume of cold water flowing from the Saint Lawrence River decreased. The flow of warm water from tropical regions into the North Atlantic began once again. There was a rapid warming in the region and the melting of the ice sheets increased. Arctic sea ice retreated northward quickly over a period of only 20 years. During this 20-year period dust concentrations in the atmosphere decreased by a factor of 3. The reduction in dust suggests a lower frequency of midlatitude storms, or a weakening of these storms. In the 20 years while the ice was retreating and in the following 30 years, Greenland temperatures warmed an astonishing 7°C. This 7° is half the difference in temperature between full glacial and interglacial climates in the North Atlantic region.

The rapid temperature drop of the Younger Dryas and the equally rapid warming after the event show that climatic change need not be a long process. Figure 6.1 shows the pattern of temperature change over the past 18,000 years.

## *THE BEGINNINGS OF AGRICULTURE*

Somewhere between 10,000 and 7000 B.C., some members of the human species changed from hunters and gatherers to agriculturalists and began to build villages and towns. During this period, domestication of plants and animals took place and the first real steps were taken toward controlling the food supply. Domestication of crops and livestock in all likelihood preceded permanent

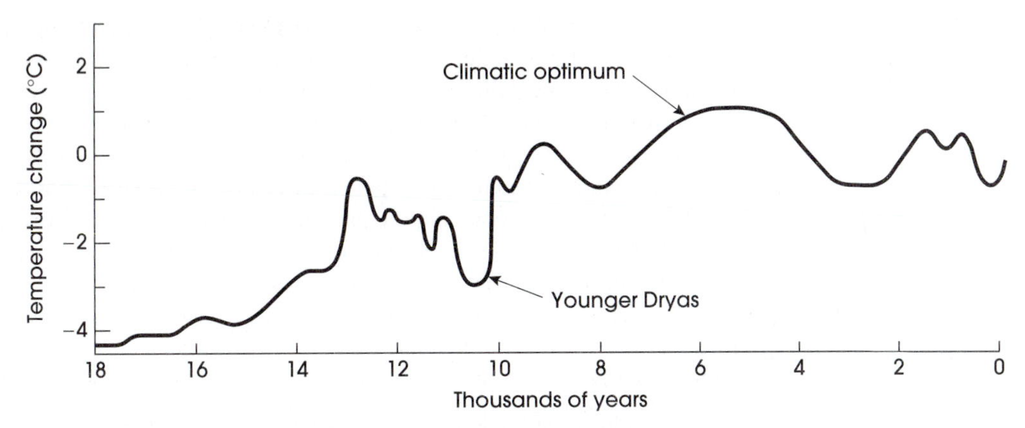


**Figure 6.1** Climate of the past 18,000 years.

settlements and was a gradual process. Humans must have learned very early to distinguish between those plants that best suited their needs for food and those that were of lesser or little use. In time, they learned they could carry plants with them in some form, prepare the ground, and plant them at a new site. By domesticating plants, they were able to begin to control the ecosystem to produce the energy requirements for their existence. Early forms of agriculture probably differed according to the regional environment. In the humid forest regions of southeastern Asia, the earliest domesticated plants were those that could be started from parts of mature plants. This is accomplished through stem cuttings or planting part of a tuber such as yam or taro.

## *FOREST AGRICULTURE*

In the forest environments, agriculture probably began on uplands in the form of slash-and-burn, or shifting, cultivation. This form of agriculture still exists in many areas and is called *ladang* in southeastern Asia, *milpa* in tropical America, and *swiddene* in other areas. The basic attributes of slash-and-burn cultivation are the same everywhere, although there are myriad variations of the basic system. The principle is to set up a temporary ecosystem of plants desirable as human food on a site cleared of forest, and then encourage forest regrowth after harvesting. The return of the forest makes it possible (at least that is the hope) to repeat the planting process at some time in the future. Initially, the brush is cleared from the planting site. Trees are limbed and girdled, if not cut down. Some weeks after clearing begins, they burn the litter. Burning reduces the bulk of the litter, but more important, it frees the mineral nutrients in the cut vegetation. This makes them soluble in rainwater and more available for the new plants. Under much of the tropical rain forest, soil nutrients are not abundant, so this extra charge of minerals is almost essential to the growth of the domesticated plants. The plots include a variety of plant species. In most slash-and-burn plots there is a structure in the domestic plants similar to that of the natural rain forest, if not of the same height. The planted species intermingle and become stratified to make maximum use of space and light. Yams, cassava, sweet potatoes, and taro grow beneath the surface. Leaves and vines of the tubers form a mat over the ground, and sugarcane, maize, or bananas grow above these. If one or another kind of plant succumbs to pests or blight, other plants provide alternative food. A variety of plants ensures a harvest in the same way that the tropical rain forest ensures its own stability over the long run. Self-sufficiency encourages a generalized agriculture to provide a varied diet.

In the Middle East there is a drier climate. Plants of the grass family with large seeds resistant to cold, heat, and drought were the first to be important for food. Wheat was probably the first crop planted in the Middle East, with barley a close second. Here, as well as in other early centers of development, farmers avoided the floodplains and high hill areas. Preference was for wooded slopes with more easily worked soils. They avoided grassland soils, with their extremely thick roots, and the higher mountain slopes, which tended to be rocky. Early agriculture was rain based. If the region had seasonal rainfall, planting was done at the onset of the rains and harvesting at the end of the rainy season. Where rainfall was more of a year-round event, planting and harvesting took place on a more even basis.

## DOMESTICATION OF ANIMALS

Initial domestication of animals must have taken place at about the same time as that of crops. In southeastern Asia, the dog was domesticated first, followed by the pig and certain fowl. The domestication of the pig could not have taken place until humans lived in settlements, because pigs are almost impossible to herd. Animals were domesticated first as much for religious reasons and pets as for food. The domestication of grazing animals began in the Middle East along with the domestication of cereal grains. These animals included the goat, sheep, ass, oxen, cow, and horse. People began using cattle as beasts of burden around 5000 B.C. Draft animals made a greater energy source available to the farmer.

Crop agriculture spread out rapidly from the centers of origin. The Danubian culture in Europe existed by 4000 B.C. The main crops were barley, wheat, beans, peas, lentils, and flax. Dogs, oxen, sheep, and pigs were kept. The settlements were mainly on the fertile, well-drained, easy-to-work soils. The plow was not yet used for cultivation. During this early agricultural stage, human beings were probably as nearly self-sufficient as at any time in the history of the species. They provided themselves with food, clothing, and shelter. The food supply was from domestic crops or livestock, or a combination of the two.

The landscape began to change from a natural one to a planned one as signs of human organization spread. Villages appeared on the landscape. The first permanent settlements depended on an assured and adequate supply of fresh water. Often, there was some economic reason for site selection, such as the existence of a valuable resource such as obsidian, chert, or flint. Early villages based on such resources existed in the crescent from the Tigris and Euphrates rivers through what is now Israel by 7000 B.C. The site of Catal Huyuk and Jarmo date from this period, and Tepe Sarab existed by 6500 B.C. The Ben Chang civilization existed in Thailand from 3600 to 250 B.C. The inhabitants were working with bronze by 3600 B.C. and smelting iron by 1500 B.C. Archeologists have recovered some 18 tons of pottery, stone, and metal items.

## DEVELOPMENT OF IRRIGATION AGRICULTURE

The development of irrigation was the next big achievement of prehistoric people. It took place independently in China, southeastern Asia, the Middle East, and South America. Irrigation must have begun on a very small scale, perhaps from a spring at the base of a hill, such as at Jericho in the Jordan Valley. It could also have begun by collecting and channeling rainwater as it ran off the hill slopes. Terrace irrigation began in different places and developed independently from riverine irrigation in the Middle East, the Philippines, East Africa, and South America (Figure 6.2).

Irrigators had some advantages over the previous slash-and-burn farmers in that they could concentrate their labor in a single area year after year. Silt and organic matter fertilized the fields,

**Figure 6.2** Primitive irrigation along the Nile River in Sudan.

and erosion was much less a problem. The fields were first cleared of obstructions such as stumps and refuse piles. Then the soil could be worked into ridges and furrows for planting and watering. The plow was invented soon after irrigation began. Ox-drawn plows were in use in Egypt and Iraq by 3000 B.C. By the middle of the second millennium B.C. plows with hoppers for planting seed were in use in Iraq. The net result of irrigation agriculture was an output per unit area far greater than from earlier rain-fed agriculture. It was more labor intensive but also more productive.

When the irrigators began to settle in the river basins, a whole new way of life began. The rich alluvial soils, along with irrigation, brought about surplus crops by providing high yields, dependable yields, and multiple crops. The rivers provided large supplies of water but demanded more labor than a single family could provide. It was necessary to construct ditches, dikes, and other facilities to carry water to the fields and to carry off wastewater. Irrigation societies came into existence through the organization of labor. Construction and management of the irrigation system demanded the voluntary or forced cooperation of a group of people. As the organization of labor for the irrigation schemes became larger, specialization of labor had to take place. A managerial class developed from the need for full-time overseers. This occupational division of labor is what gave technology its start. Specialists appeared as a response to demands from the organization. It is often said that necessity is the mother of invention. Writing was invented to keep track of the mass of laborers and supplies. Once writing began, use expanded to record history, technical discoveries, and eventually for self-expression.

The Sumerian civilization serves as a good example of a hydraulic society. The Tigris and Euphrates rivers of the Middle East start in mountain regions where there is a lot of rainfall. However, for most of their courses they flow through semiarid land. Precipitation in the highlands dissolves minerals from the soil and the streams carry the dissolved minerals and also a heavy load of silt. Annual flooding of the two rivers supplies both nutrients and water to the floodplains. The Sumerians developed irrigation between the Tigris and Euphrates rivers, and with it they prospered. Advances in technology began to take place more rapidly. By 3000 B.C., the Sumerians wrote on clay tablets. By 2500 B.C., they developed a written language that permitted literary expression. By 2000 B.C., they compiled a written history. Applied science developed with land surveying and hydraulic engineering necessary to the design and operation of the irrigation systems. Astronomy met the need to predict time of flooding. The occupational division of labor led to the definition of private land and thus to tenant farmers. The struggle for power among managers led winners to become governors and kings.

## *POLITICAL SYSTEMS EMERGE*

To manage the ever-growing populations, a sophisticated and strong central authority was necessary. This authority eventually became the regulator of all aspects of life. Religion, armies, taxes, and census takers made their appearance. These societies had a central and common focus, the waterworks. By controlling the very basis of the food supply of the populace, the ruling authority became all powerful. Civilization created a large increase in resources from the environment. With the increase in new resources came a progressively higher standard of living, particularly as measured by material wealth. The individual supposedly benefited by receiving physical security in exchange for economic slavery.

As a result of the high population densities in irrigated areas, settlement patterns changed. Villages grew into towns, and towns grew into cities. The populations of the Sumerian cities of Lagash and Unma reached 19,000 and 16,000 respectively by 2000 B.C. In fact they may have been much larger. Uruk (Ubaid) may have contained 50,000 people as early as 1500 B.C. By A.D. 37, Alexandria, Egypt, reached 1 million. For 3000 years the world's largest cities were in regions of irrigated agriculture. In A.D. 1500, Cusco, Peru, had a population of perhaps 200,000, Mexico City 300,000, and Cordoba, Spain as many as 1 million.

## *NOMADIC HERDING DEVELOPS*

As some farmers became more and more involved in grain production utilizing irrigation, they had less time to tend animals. As they cultivated more and more land, there was less land for grazing the animals. Animals grazed more and more marginal land until specialized herding societies developed. In regions where rainfall is seasonal, herders began to move livestock back and forth between lowlands and the hills, or from the river valleys to the upland pastures. Thus the pattern of seasonal migration in search of food and water came into existence. Conflict arose due to dry season competition for water and grass in the river valleys. These herders became very mobile, traveling with a minimum of personal possessions, and became acquainted with large areas of land. They became adept at warfare and hunting, and depended on meat and milk products for their sustenance. It was but a small step to becoming nomadic. This did not take place until the horse and camel were domesticated for riding about 1000 B.C. The use of horses for riding began in central Asia and spread rapidly. Nomads rode camels in the Middle East by 1100 B.C. and the practice spread across Africa in the first millennium A.D. (Figure 6.3). The success of the nomadic society depended upon movement. They grazed pastures in one area and then moved on to another. In its own way, the practice was similar to slash-and-burn agriculture. They used a portion of the grassland for part of a season and then moved on to another area. The difference is in the amount of land needed to support a family. Nomads needing from 100 to 200 times as much land as the woodland cultivator. (Figure 6.3)

In the centuries preceding the time of Christ, simple forms of agriculture spread across the grasslands of Africa. The main crops were Sudanese millet and sorghum grain. The Bantu people became very efficient at slash-and-burn agriculture and they underwent one of the most spectacular population explosions in human history. By the time of Christ there were more people living on parts of the grasslands than the land could support. The result was an out-migration of people from the Benue River valley in what is today Nigeria and Cameroon. The raw forests of the Congo proved unsuitable for this type of agriculture. The millet and sorghum were not adaptable to the hot and moist climate. Somehow, perhaps by following the tributaries of the Congo River, the Bantu managed to pass through or around the rainforest. Arriving at the Katanga Plateau south of the equator, they fanned out once more in an environment similar to that of the Sahel of Africa. Here the millet and sorghum economy thrived, and the population expanded once more. With the expansion came further geographic spreading until the people reached the Indian Ocean. They added to their basic foodstuffs the banana, Asian yam, and coco yam. All these plants arrived in Africa from Asia in the third or fourth century A.D. The Bantus had some plants that would thrive in a hot and dry climate and also some that would thrive in a hot and moist climate. Thus they could settle in virtually any environment to which they moved. By the thirteenth century and perhaps before, the Bantus had spread over much of tropical Africa. They first bypassed the rain forests and then returned to them as population pressure made it necessary.

**Figure 6.3** Nomadic herders in West Africa. They move from the desert edge into the open woodlands where they let the cattle browse on the residue of harvested fields.

Fire became an even more widely used tool than in Paleolithic times. Pastoral tribes set fires to increase the growth of new grasses for their flocks, and agricultural societies started fires to increase the yield from seed. Evidence suggests that early humans created many and perhaps all of Earth's grasslands. Trees and brush will always dominate an area to the exclusion of grass if permitted to do so. Underbrush and scrub trees take over where there is relatively little rainfall. Large trees and true forests grow where rainfall is more abundant. It does not appear that fires from natural causes are frequent enough to keep down the growth of trees and brush. The degree to which fire inhibits wood growth varies in proportion to the heat of the fire. The intensity of the fire in turn varies in proportion to the amount of fuel left by previous fires or grazing activity. Eighteenth-century travelers described the burning of the western margins of the Appalachian forests by the North American Indians. There are similar observations about primitive peoples inhabiting the grasslands of South America, Russia, Europe, and Africa.

Modern studies of the interrelationships among fire, grazing, and productivity on the African savanna show that the fire is necessary for the savanna to reach its highest productivity. A savanna is a hot, relatively dry (less than 500 millimeters of rainfall per year) plain with a cover of grasses and scattered trees. Grazing by wild animals alone is not enough to preserve a savanna as a mixture of trees and grass. Wild animal populations differ from domestic cattle, in that various species feed on different plants or on the same plant at different stages in its maturity. Because of the variety of diet, a given area with a heterogeneous cover of vegetation can support a much larger population of mixed species of wild animals than it can a population of domesticated cattle. A savanna with virtually no trees has less variety in its grass layer than one with some trees, but the variety and success of the grasses decrease when there are too many trees.

The regular burning of the savanna by tribe members permits it to maintain maximum diversity in its vegetation and in the population of game it supports. The productivity of such a savanna as a natural habitat for animals is much higher than for similar land managed by Europeans to support cattle. It is three-and-a-half times as large as that of an average virgin range supporting domestic livestock in the western United States.

## THE CLIMATIC OPTIMUM

By 7000 years ago, conditions had improved such that only remnants of ice remained. The warm period peaked about 5500 years ago. Most ice disappeared. Only the Greenland Ice Sheet and the Antarctic Ice that we have today remained. This was a time when the mean atmospheric temperature of the midlatitudes was about 2.5°C (4.5°F) above that of the present. This warm time is the *Climatic Optimum*, or thermal maximum. The term originally applied to Scandinavia when temperatures were warm enough to favor more varied flora and fauna.

In Europe temperatures averaged 2 to 3°C higher than at present. There the tree line was some 100 meters higher than today. The retreat of European ice sheets and the onset of the Climatic Optimum resulted in a pronounced drying in Sahara Desert. The drying forced many people to migrate as food supplies dwindled. Although it varied in intensity and timing, the Climatic Optimum was a global phenomenon.

## DENDROCHRONOLOGY AND DATING OF HOLOCENE EVENTS

Dendrochronology is study of growth rings in trees as a surrogate measure of changes in climate. A. E. Douglas and his colleagues at the University of Arizona pioneered tree-ring studies. Initial studies tried to relate seasonal growth of trees to sunspot cycles. In the quest for ancient living trees to study, they found the 4000 year old *Pinus aristata*. Analysis of such ancient trees permitted reconstruction of climates of the American southwest during various settlement periods.

Tree-ring analysis depends on growth rings which record significant events that happened during the life of the tree. Growth rings form in the xylem wood of the trees. Early in the season, the xylem cells are smaller and darker. The abrupt change from light-to-dark-colored rings delineates the annual increments of growth. Study of these rings, their size, and variations provides information about the varying environmental conditions in which the tree grew. The method is, of course, most valuable in determining conditions that existed in recent Holocene times. It is widely used in archeological research.

## SUMMARY

The Holocene basically corresponds to the period since deglaciation and to the time since agriculture began as a human endeavor. Rapid melting of the remainder of the continental ice sheets took place from 12,600 to 12,200 years ago, and warm ocean currents in the Northern Hemisphere spread poleward. The addition of the warm water flowing northward resulted in more glacial melting. The climate warmed rapidly, if intermittantly. The warming was interrupted by the Younger Dryas event.

Agriculture began by at least 12,000 years ago. Once begun it spread and developed into different forms depending on the regional climatic environment. Once irrigated agriculture came into being technology increased rapidly with the occupational division of labor. The climatic warming peaked about 3500 BC in what is known as the climatic optimum. By this time some form of agriculture was found in most parts of the world. Where crop agriculture was not possible, herding was practiced. In some areas land damage was already occurring due to agriculture.

## BIBLIOGRAPHY

Betencourt, J. L., T. R. Van Devender, and P. S. Martin, eds. 1990. *Packrat Middens: The Last 40,000 Years of Biotic Change*. Tucson, Ariz.: University of Arizona Press.

Fritts, H. C. 1991. *Reconstructing Large-Scale Climatic Patterns from Tree-Ring Data: A Diagnostic Analysis*. Tuscon, Ariz.: University of Arizona Press.

Turner, B. L., II, W. C. Clark, R. W. Kates, and others. 1990. *The Earth as Transformed by Human Action: Global and Regional Changes in the Biosphere over the Past 300 Years*. New York: Cambridge University Press.

chapter 7

# The Last Millenium

***CHAPTER SUMMARY***



Medieval chronicles contain many references to weather conditions. Unfortunately, although not surprisingly, most are about exceptional weather events rather than day-to-day conditions. These events include such unusual phenomena as the freezing of the Tiber River in the ninth century and the ice on the Nile River. Although these sources do not supply a continuous record, we can construct an overall view of the usual conditions by assessing the number of times that given events occurred. The freezing of the Thames River provides one such example. Between 800 and 1500, there were only one or two freezing times per century. In the sixteenth century, the river froze at least four times. In the next century it froze eight times, and there were six freezing periods in the eighteenth century. One can only suppose that progressive cooling increased the frequency of freezing.

William of Malmsbury, writing in 1125 about the Gloucester region of England, stated: "[the area] exhibits a greater number of vineyards than any other county in England, yielding abundant crops of superior quality. . . . they may also bear comparison with the growths of France." What is significant about this statement is that by the fifteenth century there was no wine industry in England. The most probable reason is a cooling of the climate since the twelfth century.

Contemporary literature from Iceland has aided in reconstructing of one of the most complete sequences of climate over the last thousand years. Data for Iceland come from the following sources:

| | |
|---|---|
| 1864 to present: | actual meteorological instruments |
| 1781–1845: | a reconstruction of weather conditions as derived from the relative severity and frequency of drift ice near Iceland |
| 1591–1780: | historical records combined with incomplete drift ice data |
| 900–1590: | information from Icelandic sagas about times of severe weather and related famines |

Such a complete record as this can be used as a base guide to the climate of the entire North Atlantic area.

## *LITTLE CLIMATIC OPTIMUM*

The Vikings settled Iceland in the ninth century.The time extending from about 950 to 1250 is known as the Little Climatic Optimum (Figure 7.1). Evidence of agriculture and other activities indicate the climate that existed at the time the Vikings settled Greenland. Eric the Red discovered Greenland in 982 A.D. In 984 A.D. the Norse founded the colony of Osterbygd on Greenland. While an icy land, it supported enough vegetation (dwarf willow, birch, bushberries, pastureland) for settlement. The settlers brought cattle and sheep that not only survived but thrived for a considerable period. They established two colonies and began to farm. The outposts thrived and regular communications existed with Iceland.

Between A.D. 1250 and 1450 climate deteriorated over wide areas. Iceland's population declined. Grains that grew there in the tenth century would no longer grow there (Figure 7.1). Greenland became isolated from outside contact, with extensive drift ice preventing ships from reaching the settlements.

In Europe storminess resulted in the formation of the Zuider Zee, and the excessively wet, damp conditions led to a high incidence of the disease St. Anthony's fire (ergotism).

Research shows that Indians in the Great Plains migrated as a result of variation in precipitation. A study by archaeologists and climatologists showed, for example, that Indians of the Mill Creek culture of Iowa deserted a thriving community about A.D. 1200 when precipitation declined rapidly.

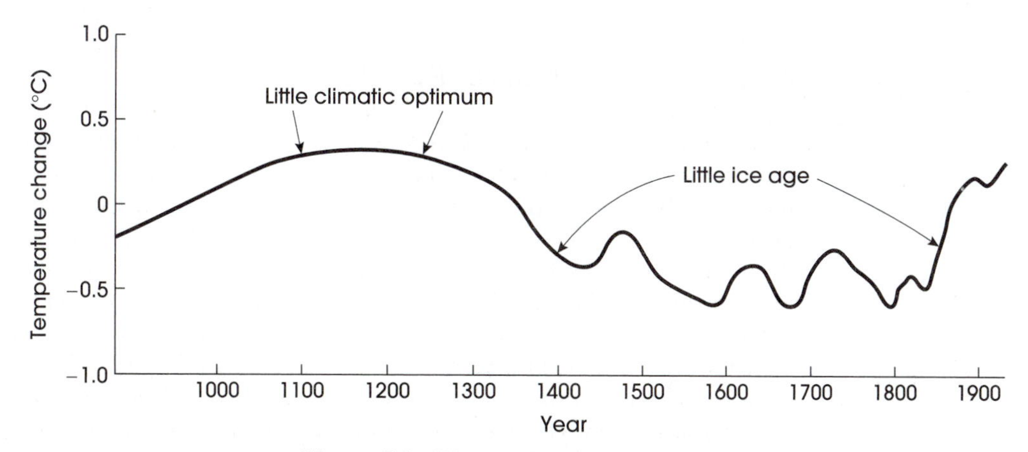


**Figure 7.1** Climate of the last millenium.

The impact of severe drought on settlements has been used to reconstruct past climates. As an example, the abandonment of settlements in the southwestern United States probably was due to drought. The abandoned settlements of Chaco Canyon and Mesa Verde show that they supported large and prosperous communities. By A.D. 1300, these settlements were deserted. Reasons other than drought may account for the abandonment. However, tree-ring analysis shows that between A.D. 1276 and 1299, practically no rain fell in these areas. Extensive migration of people resulted from the drought.

## *GREENLAND SETTLEMENTS AND THE LITTLE ICE AGE*

The Little Ice Age was the coldest period in historic times. It occurred from the fifteenth to the nineteenth centuries. During this period, areas bordering the North Atlantic Ocean experienced drastic cooling. Mountain glaciers expanded and in some cases reached their maximum extent since the end of the Pleistocene glaciation. Three major periods of expansion of glaciers took place in the years 1600 to 1650, 1810 to 1820, and 1850 to 1860.

The little ice age marked the end of the Norse settlements in Greenland that had begun in the tenth century. In fact, in 1492 the Pope complained that none of his bishops had visited the Greenland outpost for 80 years. (He was not aware that the settlements were already gone.) Ice in the northern seas prevented traffic from reaching Greenland. By 1516 the settlements had practically been forgotten. In 1540 a voyager reported seeing signs of the settlements but no signs of life. The settlers had perished.

After flourishing for more than 400 years, the colonies disappeared about A.D. 1410. Excavations show that at first the soil permitted burying bodies at considerable depth. Later graves became progressively shallower. A Danish archeological expedition to the sites in 1921 found evidence that deteriorating climate must have played a role in the population's demise. Some graves were in permafrost that had formed since the burials. Tree roots entangled in the coffins showed that the graves were not originally in frozen ground. It also showed the permafrost had moved progressively higher. Examination of skeletons showed that there was not enough food. Most remains were deformed or dwarfed. There was clear evidence of rickets. All the evidence points to a climate that grew progressively cooler, leading eventually to the settlers' isolation and extinction. Now the soil in the area of the Norse colonies is frozen all year. It is not certain that the colonies failed due to climatic reasons, but it seems likely.

## *NORTH AMERICAN COLONIES IN THE LITTLE ICE AGE*

The colonies in eastern United States suffered from the cold of the Little Ice Age. The soldiers of the American Revolution suffered in the cold weather. Sometimes the unusual ice served as a useful tool. British troops, for example, slid their canon across the frozen river from Manhattan to Staten Island.

The year 1816 is known as "the year without a summer." The year began with excessively low temperatures across much of the eastern seaboard. When spring came, the weather was cool but not excessively so. In May, however, the temperatures plunged. In New England, frost occurred in every month of the year.

In Indiana there was snow or sleet for 17 days in May. This killed off seedlings before they had a chance to grow. The cold weather continued in June, when snow again fell, devastating any remaining budding crops. No crops grew north of a line between the Ohio and Potomac rivers, and returns were scanty south of this line. In the pioneer areas of Indiana and Illinois, settlers had to rely on fishing and hunting for food. Reports suggest that raccoons, groundhogs, and the easily trapped passenger pigeons were a major source of food. The settlers also collected many edible wild plants, which proved hardier than cultivated crops.

It was cold in Europe also. Alpine glaciers grew in size and advanced to lower elevations. The Thames River in England froze over many times. It has not frozen over since the winter of 1813–1814.

## *THE ERUPTION OF KRAKATOA IN 1883*

Plate tectonics is one of the long-term forms of change on the planet. The slowly moving plates react to the internal heat in a variety of ways, as we discussed in Chapter 2. Volcanic eruptions are always occurring at some place on the planet. One of the largest eruptions of modern times is that of Krakatoa.

Krakatoa is the name of both an individual island and a small group of islands in the Sundra Strait, between Java and Sumatra. The islands in the group in 1883 were Krakatoa, the largest, Verlaten, Lang, and Polish Hat. Krakatoa itself was about 8 kilometers long and 3 kilometers wide with three volcanic cones. The highest was Rakata, with a height of 810 meters, and the others were Danan (445 meters) and Perboewatan (122 meters). The last known eruption on the islands had been in 1680 and the volcanoes were considered extinct.

In mid-May of 1883 the islands of Krakatoa began to show renewed volcanic activity. The first eruption was from Perboewatan. The island group was uninhabited at the time but large numbers of people lived on surrounding islands, primarily on the shorelines of Sumatra and Java. A group sent from Batavia, Netherlands East Indies (now Jakarta), went ashore and climbed to the top of the crater. Boiling lava was visible in the cone and thick deposits of ash and pumice were found to cover some parts of the island. Activity continued on the island in intermittent form into the month of August. Residents of nearby villages and the crews of ships passing through the straits reported numerous episodes of explosions, ash fall, and earthquakes emanating from Krakatoa from May through August. On August 11 smoke was visible coming not only from the cones on the island but from a number of different points on the islands as well.

At 1:00 P.M. on August 26, the main eruption began and lasted about 24 hours. There are no accurate visual reports of what actually happened on the island during this time. Few people with a line of sight to the island survived the ensuing events, and due to heavy ash falls from the volcano the island was obscured by almost total darkness for a period of three days. The northern part of the island erupted violently through the remainder of the day. The ash was so thick that darkness fell over a considerable distance from the island. Explosive eruptions occurred throughout the afternoon of August 26 and lasted throughout the night. One of these was so strong that it was heard at Daly Springs, Australia, 3200 kilometers from Krakatoa. Several low-lying villages were hit by tsunamis, destroying buildings and drowning hundreds of residents. Survivors were later to report heavy electrical activity in the vicinity of Krakatoa, described as "balls of white fire" and "chains of fire." Crews of ships in the strait reported hot sulfurous winds and warm seawater. Early the following morning, the paroxysmal eruption reached its peak. Tsunamis exceeding 40 meters in height devastated villages on Sumatra and Java. The most explosive blast of all took place at 10:02 A.M. The sound of the explosion was heard at least 3300 kilometers in all directions, and it was heard at Rodriguez, 4653 kilometers to the southwest. The explosion blew away the northern three-fourths of the island, producing a caldera 10 kilometers across and over 300 meters deep. The pressure waves from the blast traveled around the earth at least seven times.

It is estimated that 72.9 cubic kilometers of earth material was blown into the atmosphere. The island of Polish Hat disappeared completely, and the islands of Lang and Verlaten grew in size, largely due to the mass of ash and rock deposited from the eruption. Most of the ejected material was deposited over an area about 60 kilometers in circumference. The eruption was the most explosive eruption of modern times.

The eruption of the volcano by itself did not do a tremendous amount of damage or cause loss of life beyond that on the island group itself. The lethal and damaging part of the event was the tsunamis generated by the concussions. The first took place on the evening of August 26. The highest were produced by the shock of the tremendous blast at 10:02 A.M. on August 27. The waves from the l0:02 blast reached more than 40 meters high at Angier, Java. The tsunamis destroyed hundreds of coastal villages on islands with shores facing Krakatoa. Among the villages and towns destroyed were Anjier and Merak on Java and Katimbang and Telok Betong on Sumatra. Katimbang was almost entirely destroyed by a tsunami that hit at 8:30 P.M. on August 26. Only the district hall, built on a high terrace, survived; over 8000 residents were killed. Telok Betong was hit by a lesser wave several hours later, then destroyed the following morning by higher waves. The densely populated west coast of Java received most of its damage during the early morning hours of August 27. Only a few persons survived an immense wave that struck Anjier and Merak soon after 6:00 A.M. The death toll in this district exceeded 20,000. Later waves were even higher, but the coastal towns had already been destroyed.

The power of these tsunamis is perhaps most dramatically evidenced by the single block of coral weighing over 600 metric tons that was thrown ashore near Anjier. At Telok Betong, a steamer was washed 2 miles inland and its crew of 28 killed. Many deaths from burns were reported in Sumatra, attributable to hot tephra falling like rain. Even at Batavia, 160 kilometers from Krakatoa, extensive damage and casualties from tsunamis were reported. More than 36,000 persons perished due to injuries or drowning.

The tremendous explosion on the morning of August 27 largely ended the eruption. On August 29 visibility had reached the point where it could be discerned that the island was largely gone and volcanic activity reduced to insignificance.

Indirect effects of the eruption of Krakatoa were noticeable around the world. The tremendous volume of fine ash and aerosols erupted into the stratosphere, produced spectacular sunsets around the world and lowered global average temperatures by approximately 0.5°C over the following year.

## *THE TUNGUSKA EVENT OF 1908*

On the morning of June 30, 1908, an extraterrestrial object entered the earth's atmosphere. The trajectory was generally south to north over Asia. When the object dropped to an altitude of about 135 kilometers, friction with the atmosphere heated it to incandescence. The cylindrically shaped intruder glowed with a brilliant bluish-white color and began to stream a trail of debris as it moved northward over Russia. Glowing brighter than the early morning sun, it was visible to people over thousands of square kilometers for a period of about ll minutes. At 7:10 A.M. local time, the object, probably less than 40 meters in diamter with a mass of some 50,000 metric tons, exploded at a height of 3 to 8 kilometers over latitude 60° 55″N and longitude 100°E (Figure 7.2). Ground zero was a large marshy basin called the Southern Swamp near the Stoney Tunguska River in the Taiga of Siberia, some 65 kilometers north of Vanovaro.

The large body did not strike the earth as a unit. Trees were burned but left standing directly under the point of ignition. Many small circular ponds 10 to 200 meters in diameter have been

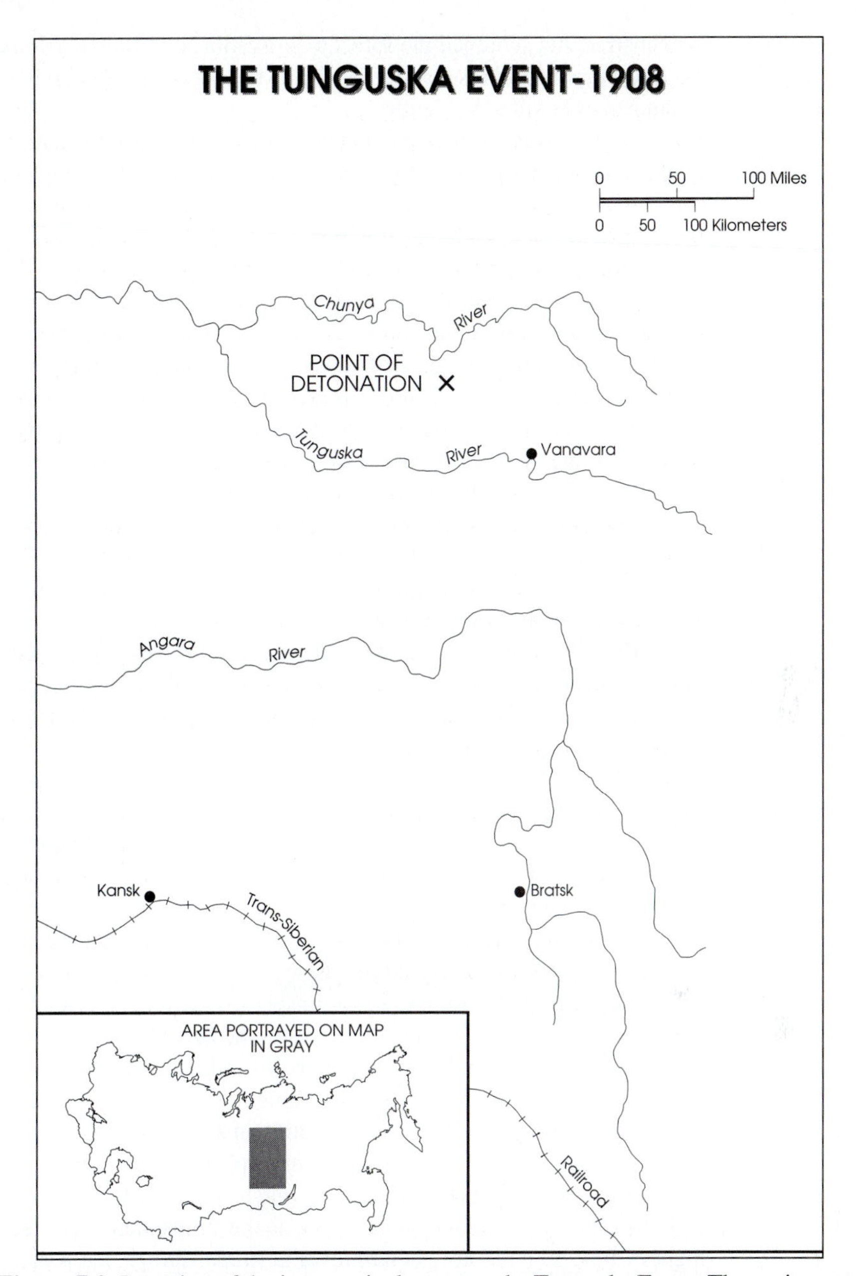


**Figure 7.2** Location of the impact site known as the Tunguska Event. The region was relatively uninhabited at the time. There was tremendous environmental damage to forests, but there was no known loss of life.

found on the ground beneath the point where the object exploded, but no meteoric or other blocks of material were ever found.

The detonation was first marked by a tremendous flash of light and heat, with as much as 30 percent of the total energy released as radiant energy. The intense heat melted the permafrost to an unknown depth, causing flooding in the local area and even on some large rivers. The heat also charred and scorched the forest over a radius of 13 to 16 kilometers. The instant ignition of the forest produced fires that burned for up to a week. All animal life within an area of 260 square kilometers was killed by the heat.

The area was thinly populated by a group of people known as the Tungas. They were a semi-nomadic people living from the products of reindeer herds and by hunting and trapping. The person living closest to the ignition center was a Tunga named Vasily Ilich. He owned a herd of 1500 reindeer in the area, and had constructed a number of small wooden buildings which were used largely for the storage of goods. All were incinerated or melted. Only a few items of hardware were recognizable, and the reindeer herd had perished. The concussion of the explosion was tremendous and caused the earth's crust to shake. In Kansk, 600 kilometers to the south, people felt the ground move and hanging objects swung back and forth while utensils fell from shelves. The concussion was such that it was recorded on seismographs more than 5000 kilometers away and severe atmospheric shock waves were produced by the explosion. Trees were flattened outward in all directions from the swamp, with upright but seared forest in the center, since named "the telegraph forest" for the resemblance of the trees, stripped of their limbs, to telegraph poles found along railroads. Farther away from the center the ridges were all blown clear of vegetation, with the scorched tree trunks all lying facing the same direction. The forest survived only in the deeper valleys, which were protected from the flash of heat and atmospheric shock waves. In all, an estimated 80 million trees were leveled by the blast.

There were at least two atmospheric shock waves, which became increasingly far apart in time as distance from the site increased. Several small villages were flattened by the shock waves, and windows were broken 80 kilometers away. At Kansk, 600 kilometers away from the blast site, people and animals were knocked to the ground. The second shock wave at Kansk followed the first by 5 to 7 minutes and was stronger than the first wave. Many people were injured, primarily by the atmospheric waves. Some were knocked off their feet, some rendered unconscious, and one person is known to have been deafened at a distance of 40 kilometers from the blast site. The sound of the concussion was heard at Turkoyansk, 1000 kilometers to the west. The pressure waves in the atmosphere traveled around the world twice, and produced wide-ranging fluctuations in pressure. In England these oscillations took place some five hours after the explosion occurred, and lasted for nearly 20 minutes.

A huge fireball rose into the atmosphere reaching an altitude of more than 20 km and was visible for hundreds of kilometers away. As the fireball expanded it sucked in air that produced a windstorm on the ground that picked up dirt, ash, and debris. Dark masses of cloud towered upward replacing the fireball. The mass cooled as it rose and with the ash and dust serving as nuclei, water condensed in the cloud resulting in a black rain falling in the vicinity.

The explosion produced strange atmospheric effects for days afterward, and the phenomenon were most noticeable at night. Over much of Europe the northern sky was bright on the nights of June 30 and July 1. High white clouds in the atmosphere reflected so much sunlight that on the night of June 30 it did not get dark over northern Europe and Asia. Sunset around the world was spectacular for several days as the dust cloud spread. The light was visible throughout the night, and was white to light yellow in color. The nightly displays continued until July 3. Analysis in recent years indicates that not only was the atmosphere disturbed, but so were the earth's magnetic and gravitational fields.

Estimates as to the magnitude of the explosion vary greatly. There is no doubt that it was of a size not experienced previously in historic time. It has been estimated by well-known scientists using established procedures to have been as much as 1500 times greater than the bomb dropped on Hiroshima, Japan, in July 1945. It has not yet been established for certain what the object was that exploded. It has been defined as being a meteorite, a comet, a spaceship, and even a black hole. Whether it will ever be established for certain exactly what the object was is debatable. That it came from space is certain.

The impact of this intruding object on human life and property was amazingly small. It exploded over one of the least inhabited areas of the earth's landmasses. Had the object exploded over the ocean, huge tsunamis would have been generated that would have been devastating. As it was, it exploded over the taiga, destroying forests, wildlife, and domestic reindeer, injuring a number of people, but as far as is known, took no human lives.

## *COLD AND SNOW IN THE LATE 1970S*

A concern that rapid cooling might herald a return to the ice ages was heightened by a series of severe winters beginning in 1976–1977. During the four winters of 1974–1975 through 1977–1978 there was a most unusual variety of weather patterns with extreme departures from normal. The weather was noteworthy over the Ohio River Valley states of Illinois, Indiana, Ohio, and Kentucky. The most dramatic changes took place aloft in the atmosphere at the 700-millibar level. January 1975 temperatures in the Ohio Valley were 1.7 to 2.2°F above the normal of 1941–1970. Precipitation amounts averaged 25 to 40 millimeters above average, and of this only about 10 percent came as snowfall.

In the winter of 1975–1976 the upper air circulation continued to change. That winter was the first of three consecutive winters with temperatures well below the normal of 1941–1970. Temperatures averaged 4 to 5°F below those of the preceding winter. However, this was only a prelude to the record-setting events of the next two winters.

### The Cold Winter of 1976–1977

January 1977 proved to be the coldest month on record in the Ohio Valley. Circulation features changed radically from those of the preceding January. This circulation had three main features:

1. A very deep Aleutian low that expanded to cover nearly all of the northern Pacific Ocean
2. A large ridge of high pressure over the west coast of North America
3. A very deep low over Quebec Province that was south of its usual location

This pattern of circulation set in early in the fall of 1976 and continued with little change until the end of February 1977.

The development of both the Pacific coast ridge and the trough over the eastern part of the continent altered the flow at the 700-millibar level. The 700-millibar height departures exceeded +90 meters along the west coast of Canada between Washington and Alaska. They were −120 meters over New England. The 700-millibar winds in the Ohio Valley had a strong northerly flow, bringing cold arctic air into the valley. Wind speeds at 700 millibar were as much as 8 meters per second above normal off the south Atlantic coast of the United States.

In October 1976 cold airstreams poured southward, setting records for cold temperatures. Cities in 10 states, mostly in the southeast, recorded the coldest October on record, some cities recording temperatures as much as 5°C below normal. Temperatures north of the Ohio River dropped as much as 11°C below the January normal. Indianapolis, Indiana had below-freezing temperatures continuously from December 28, 1976 through January 1977. Detroit, Michigan and Buffalo, New York also recorded temperatures continuously below freezing for January.

The effects of the cold weather of 1976–1977 were extensive and varied. The problem was as much a result of the persistence of the cold as it was a result of the absolute temperatures. The three-month period from November to January was the coldest in 105 years. Many deaths resulted from the cold. Some people froze to death in their homes when heating systems failed or could not provide enough heat. The elderly and people living alone in rural areas suffered the most. Rural people often had no means of communication to alert others to problems. Other people froze to death in automobiles or froze trying to walk to safety when their vehicles stalled on rural roads.

Public water supplies stopped working because water lines froze. Many pipes were not deep enough underground to escape the frost. Schools and factories closed due to lack of fuel. Pumps could not move natural gas through the lines fast enough to keep up with demand. Transportation systems failed. O'Hare Airport in Chicago shut down overnight near the end of January because of the cold. The continuous cold took much ground equipment out of service. Freezing weather along the Gulf coast damaged citrus fruit and other crops.

The high-pressure ridge over the west coast very effectively blocked out the moisture from the western half of the continent. To the east of the Rocky Mountains the strong northerly flow of air kept the moist air from penetrating the continent from the south. Much of the continent was well below normal for precipitation in November, December, and January. Extreme drought developed in the far west.

Although precipitation generally decreased, snowfall increased. Lake-effect storms increased and produced extreme snowfalls at some cities. Hooker, New York, recorded 11.8 meters of snow for the 1976–1977 winter. In the winter of 1976–1977 there were 51 days with lake-effect snowstorms, more than twice the average of the previous two winters. The storms began in October and the last one occurred in February. The worst event was the five-day storm beginning on January 28. For the first time in the United States, the federal government declared several counties disaster areas based on snow depth alone. Wind produced drifts as high as 10 m.

Northern areas were not the only ones affected. On January 19, 1977, it snowed in Miami, Florida and as far south as Homestead, Florida at 25°35′N latitude. The area of North America covered by snow was at a record extent in December 1976. About 65 percent of North America was snow covered, and in January 1977 even more of the continent was snow covered.

## The Snowy Winter of 1977–1978

January 1978 saw the northward displacement of the low pressure over Canada to a position over Baffin Island. This northern migration resulted in a reduction of the north–south height gradient over the eastern United States compared to January 1977. The major trough in the eastern United States remained shallow. Winds over the midwest continued to have a strong northerly flow, but not as strong as in January 1977. The major ridge in the western United States had a more inland position compared to the previous year. The 700-millibar height departures were nearly −50 meters during the previous winter.

In January 1978, most of the United States had above-normal precipitation. Some areas had as much as three times the normal precipitation. Due to the cold temperatures, much of this precipitation fell as snow. Stations scattered throughout the midwest received more than 750 millimeters of snowfall. Chicago received 1.9 meters and South Bend, Indiana more than 4 meters for the three-month period December through February (Figure 7.3).

During February 1978, precipitation totals were low. In Indianapolis precipitation for the month was the second lowest on record. The continued cold temperatures of February kept the January snowfall from melting and additional snow fell in February in many areas. The area covered by snow in mid-February set a record (Figure 7.4). Snow covered 17 $\frac{6}{10}$ million square kilometers

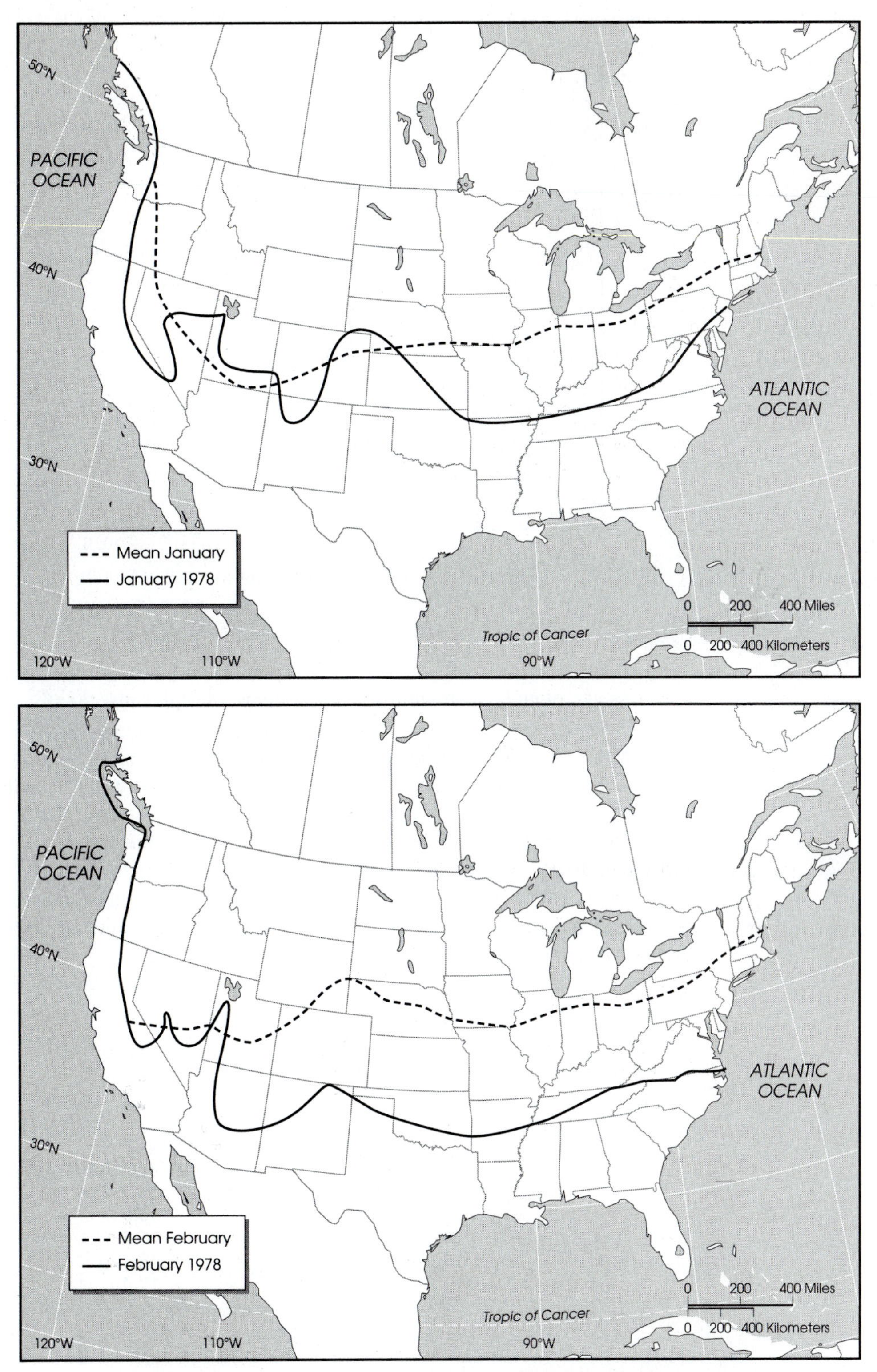


**Figure 7.3** Limit of snowfall in January and February 1978 and 1979.

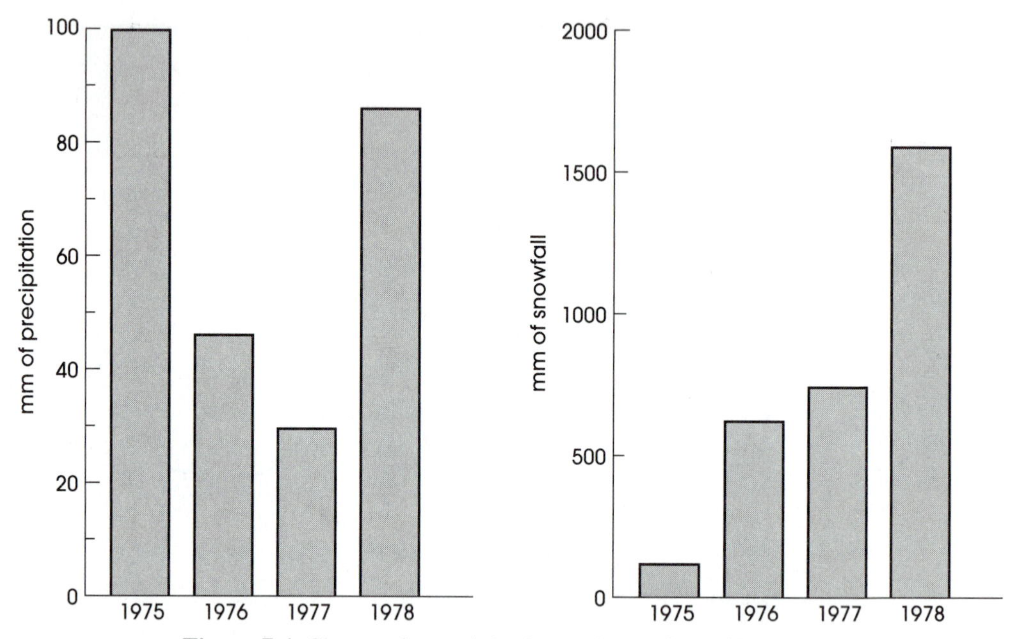


**Figure 7.4** Changes in precipitation and snowfall, 1975–1978.

of the continent at the time. This was 500,000 square kilometers more than in January 1977 when the previous record was set. Table 7.1 provides the data for snow coverage during these winters.

As can often happen with weather systems, change came rapidly. Cold temperatures and snow ended abruptly in March. Stations in western states reported record or near-record high temperatures for the month. Eastern states that had so much snow on the ground at the beginning of the month escaped major flooding only because precipitation dropped well below normal during March.

The temperature and snowfall extremes that characterized these winters generated an explosion of debate over the chance that Earth was returning to the ice ages. Many books appeared heralding the return of the ice ages. Hundreds of articles in popular and professional journals also predicted a plunge into another ice age. As time has passed it appears this is not the case, at least for the remainder of the twentieth century.

**TABLE 7.1** Area of North America Covered by Snow in Mid-January in the Years 1975–1978

| *Year* | *Area (millions of square kilometers)* |
|---|---|
| 1975 | 16.2 |
| 1976 | 16.2 |
| 1977 | 17.1 |
| 1978 | 17.4 |

## OTHER COLD TIMES IN THE TWENTIETH CENTURY

Other cold winters occurred before during historic times. Lofgren and Fritts, using tree-ring data, reconstructed yearly pressure patterns for the years 1600 to 1899. They found that pressure patterns similar to those of 1976–1977 occur rather frequently for several decades and then disappear for several decades.

The 1917–1918 winter was the coldest winter on record for the Middle Atlantic and New England states. There are some similarities between the winter of 1917–1918 and 1976–1977. In each case the fall months were cold over most of the eastern side of the continent. Temperatures in October 1917 were still the lowest on record for that month, although October 1976 was among the coldest. In 1917–1918, the prolonged cold led to fuel shortages and distribution problems, as in 1976–1977. In both years January was extremely cold but February was mild.

Winter circulation patterns similar to those of 1976–1977 occurred in four other winters during the decade of the 1960s. They are the winters of 1960–1961, 1962–1963, 1967–1968, and 1969–1970.

## RECENT GLOBAL WARMING

Fortunately, by the end of the nineteenth century, the instrumental record shows that the climate was again improving. A reconstructed record of temperature is shown in Figure 7.5. If we consider 1950–1980 to be the baseline period, we see that the years before were cool but warming, while those after were even warmer. It is the latter trend that raises the specter of global warming (Chapters 18 and 19). Beginning about 1875 the world climate became warmer and wetter. It is one of the better periods in Earth's recent history for life on the planet, certainly the best conditions in the last 1000 years. Earth more than doubled its human population in this period, and it may be these more favorable conditions that permitted the population to expand so rapidly. The most rapid warming took place from about 1885 to 1940. The temperature increased a little more than 1°C in the winter and $\frac{1}{2}$°C in summer.

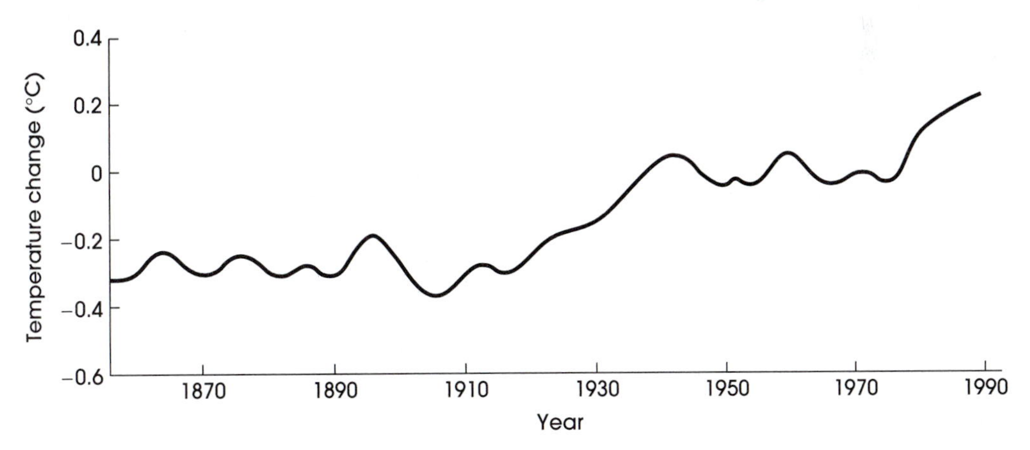


**Figure 7.5** Temperatures from 1870 to 1990 expressed as departures from the climatic normals of 1951–1980.

The largest changes appeared in the northern hemisphere, where winter temperatures over the arctic increased 3.3°C from 1917 to 1937. In Greenland, Iceland, and Scandinavia, temperatures rose 3.9°C above those of 1900. The average winter temperatures at Spitzbergen increased 10°C in 40 years.

Earth's mean temperature peaked in the 1940s and began to cool. The amount of cooling and the time of onset of the cooling varied from place to place. Polar regions and midlatitude deserts showed the most cooling. Mean temperatures in parts of the arctic dropped 3°C. A drop of 0.4°C reduced the growing season in England by more than one week. Fishing fleets in Iceland that had ranged farther north over the last century retreated southward. For the first time in the twentieth century drifting ice appeared in the vicinity of Icelandic ports. On June 2, 1977 the 8272-ton ferry *William Corson* sank after striking an iceberg off the coast of Newfoundland. Contrary to the sad events associated with the sinking of the *Titanic*, the 85 crew members and 41 passengers all abandoned the vessel safely and were rescued. No comparable cooling took place in the southern hemisphere.

## SUMMARY

Over the last 1000 years there have been many fluctuations in Earth's climate. The fluctuations have not been nearly as large as those that took place over longer periods. Much more information is available, including written documents and more recently, direct measurement of global change.

At the beginning of the millenium the Little Climatic Optimum occurred. Weather was unusually warm for several centuries. Human settlement spread toward the arctic. Iceland and Greenland were settled, as were other islands in the North Atlantic Ocean.

Many of the oscillations over this time lasted but a few years. Others persisted for centuries. Interannual variations in Earth's climate have ever-greater impact on the human species. Since the human population has grown rapidly during this time, changes of the same size in the twentieth century will stress more people than in the tenth century. Times of extreme cold and warmth cause extreme hardship in affected areas. Times of abnormal amounts of precipitation cause equally large social disturbance.

chapter 8

# Drought, Famine, and Climatic Oscillation

***CHAPTER SUMMARY***



The general circulation of the atmosphere changes from year to year. When it does change it brings changes in the distribution of precipitation. Some areas then receive more rainfall and others less rainfall. When more rain than normal occurs, floods result. When rainfall declines, drought sets in.

Much of Asia, the Middle East, and equatorial Africa are subject to the effects of an atmospheric circulation system known as the monsoon, and herein lies part of the explanation for the frequent and extensive droughts and associated famines that have taken place. Figure 8.1 delineates the region of Africa and Asia affected by monsoons. A monsoon is a very pronounced seasonal shift in atmospheric pressure, which in turn brings about a marked change in wind direction, atmospheric humidity, and precipitation.

During the summer months, convergence predominates over Asia and high humidity and precipitation are general over the continent. In summer the low-pressure system and convergence of air are strong and the onshore flow of moisture-laden air blowing north from the Indian Ocean and the Arabian Sea is considerable. The winds converge over the northern plains and slopes of the Himalayan Mountains, a region that exhibits one of the highest amounts of rainfall on the

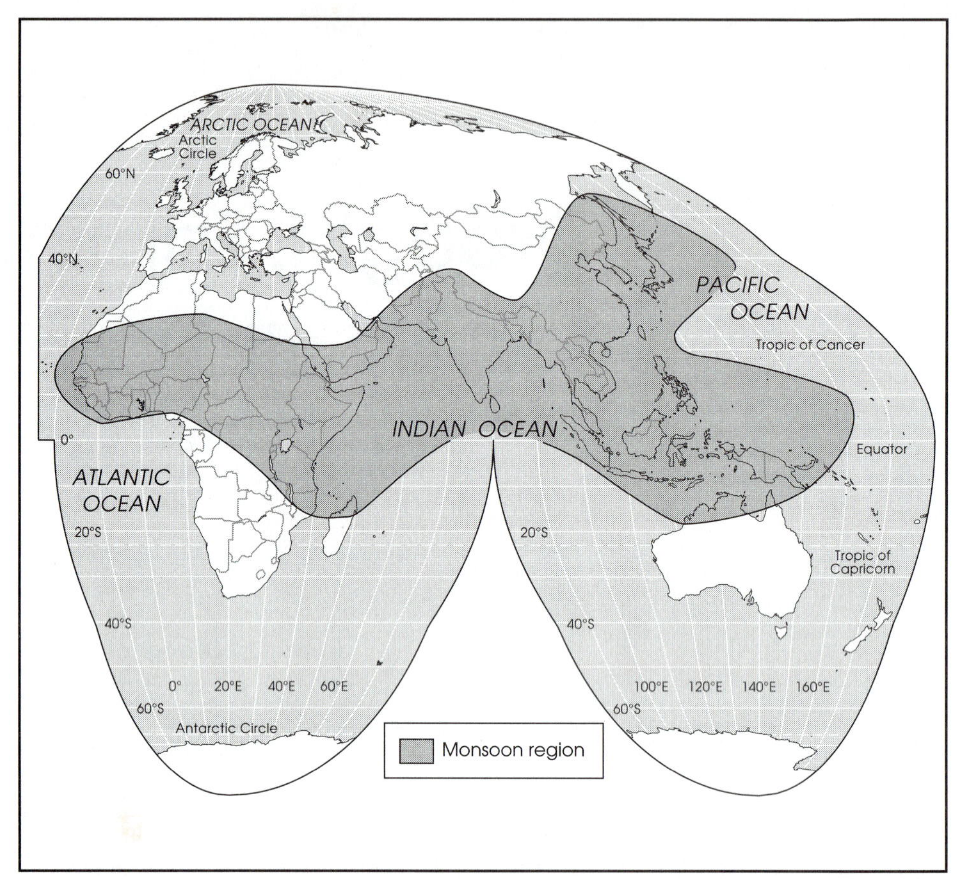


**Figure 8.1** Monsoon region of Africa and Asia. Within this extensive area, seasonal reversals of wind direction are common and pronounced.

planet. Cherripunji, India, is often used as an extreme example of seasonal rainfall, where the annual precipitation has reached 25 meters, with most of the rain falling during a four-month period. The resulting precipitation on the coastal areas and oceanic sides of offshore islands is substantial.

The summer monsoon winds blow across the subcontinent of India from southwest to northeast from April to October. The winds reaching the west coast of India have traveled over the ocean and have a very high moisture content. As the winds move onshore they produce a great deal of precipitation. At Bombay the rains last from June until the end of September and total over 2 meters. The winds pass over the western Ghats Mountains across the plains of Bengal toward the Himalayan Mountains. As a result of the directional character of the winds and the topography of the Indian subcontinent, the annual rainfall varies tremendously. The western Ghats divide India into two distinct climatic zones, as they run nearly perpendicular to the direction of the summer winds. There is a narrow, very wet climatic zone on the western side of the Ghats and a much wider and drier zone on the eastern side of the mountains.

During the winter months, when high pressure dominates the Asian landmass, divergence is more frequent than convergence. Divergence from this strong system north of the Himalayas produces extremely strong offshore surface winds. South of the Himalayas and south of the jetstream, the anticyclonic flow is much weaker, although offshore flow prevails. When divergence is present, subsiding dry air covers much of the continent, resulting in lower humidity and less precipitation. The subsidence produces offshore winds along much of coastal Asia. As the air streams out over the ocean, it evaporates large amounts of water from the sea surface and thus becomes a source of moisture for the offshore islands and along some coastal areas. There is a secondary rainy season in parts of India with the offshore flow of air, and a second-season harvest in March and April.

The monsoon is the product of a variety of factors working together, including migration of the intertropical convergence zone (ITCZ). The intertropical convergence zone is the dividing zone between the northeast trades and the southeast trades along which convergence frequently occurs. The ITCZ migrates north to between 25 and 30° N. The land mass interior records a large change in pressure that enforces the migration of the ITCZ; and it is likely that the Himalayan Mountains are a very significant topographic barrier that splits the atmospheric circulation over Asia into two subzones, one over the main Asian landmass, and the other over the Indian subcontinent and adjacent areas. In the summer, as the temperature differential over the continent diminishes from south to north, the subtropical jetstream breaks down and the subtropical high and ITCZ shift rapidly northward. This brings a sudden shift of winds onshore over the Indian subcontinent. At the same time, north of the Himalayas, a weaker convergent system develops with attendant onshore flow of air. The precipitation is primarily convectional rainfall that is randomly distributed. Some traveling low-pressure storms associated with the ITCZ add additional rainfall. The peak season of rainfall moves northward over India as the zone of convergence moves northward toward the slopes of the Himalayas.

The Asian monsoon is quite variable from year to year in terms of the time of year when the onshore and offshore flow begins and ends and in terms of the duration and intensity of the rainy season. The total amount of rainfall varies as a consequence and the agricultural economy of northern India and Pakistan are cruelly subject to the whims of the system.

It is this monsoon that is the most important for most of India. The primary growing season for rain-fed crops is during the summer rainy season, with the main harvest in December.

## IMPACT OF DROUGHT IN THE DEVELOPING COUNTRIES

Drought has a much greater impact on people in the developing countries than it has on people in industrial societies. The primary reason for this is that in the developing countries there is more dependence on agriculture as a way of life. When crops fail or there is not enough forage for livestock, there is an immediate effect on the populace.

The demand and supply of food have been in a delicate balance for the human species throughout history. When the food supply has increased, there has been a gain in population, and when food has been in short supply, some sort of trauma has been inflicted on the populace. Starvation results from insufficient food intake. During the long period of the hunting and gathering societies, starvation was probably near at hand for individuals, family groups, and tribes. The development of agriculture allowed the world population to expand rapidly and in large numbers. At the same time, the basis for the supply of food—agriculture—became more directly dependent on the weather. Famine, a phenomenon that affects large numbers of people over a broad area, did not become a part of human experience until after agriculture began. However, as agriculture expanded, so did the frequency of famines. The number of times that famine has spread over the land is high. All histories of peoples and nations record famines. There are repeated references to famine in the Bible and other ancient histories.

## *DROUGHT AND FAMINE*

There are many causes of famine, but one of the major ones is drought. Most catastrophic famines have been precipitated by drought. Drought affects the quality and quantity of crop yields and the food supply for domestic animals. In the case of severe drought, there may be a great loss of domestic animals. The loss of milk products or meat itself precipitates the effect of the drought. Great famines have occurred throughout the Asian continent from the time that agriculture spread over the continent. India, China, Russia, and the countries of the Middle East have all suffered many times from drought-related famine. A famine is described as occurring during the time of Abraham (about 2247 B.C). Another massive famine occurred in Egypt prior to the exodus of the Israelites. Drought and famine are endemic in India and China. The oldest record of famine in India goes back to 400 B.C. and in China to 108 B.C. From the time of the earliest known famine there have been nearly continuous episodes of drought and famine in some parts of Asia.

## *DROUGHT AND FAMINE IN INDIA*

In the mid-eighteenth century the people of India were largely subsistence farmers. It was primitive crop agriculture subject to the vagaries of the monsoons. The country was under the control of the British East India Company, which kept the farmers on the verge of starvation under the best of conditions. Because of the general poverty of the mass of the population and the marginality of the food supply, only a small shortfall produced scattered starvation. India is a huge country, and at the time only a ponderous transportation system existed. There was no means of moving large quantities of food, nor of moving people to more productive areas. Therefore, when drought set in, the alternatives were few.

The beginnings of a prolonged drought and massive famine began in India in the fall months of 1768. Rainfall was below normal and the crops were poor in December. The summer monsoons did not produce the usual rain in 1769, and again the crop yields were scanty. The drought persisted and there was little rain in the fall of 1769, with October virtually without rainfall. There was little precipitation in the spring of 1770. It was not until May that significant rain occurred.

Famine was apparent in the northern parts of Bengal by November 1769. By April 1770 over 30 million people in West Bengal and Bihar were affected. Estimates of deaths ranged upward to 10 million. The deaths were due to a combination of starvation and disease. Smallpox became epidemic in association with the drought. The death toll from this famine is the greatest known up to this time resulting from drought and famine. The death toll was exacerbated by the flood of people that fled the countryside and moved into the cities looking for sustenance—and there was none to be found.

In the years immediately following the famine there was a public outcry in Britain, resulting in a number of parliamentary acts to improve the economic situation of the Indians and to plan drought relief. Some improvements were made, but not nearly enough.

## *DROUGHT AND FAMINE IN ASIA, 1875–1879*

It was just a little more than a century after drought and famine struck India that the process was repeated with even more lethal results. This drought was intense, prolonged, and covered much of Asia. The best documentation regarding the event is for India and China, but without a doubt other regions were affected. By this time in history better records were being kept on environmental parameters such as rainfall, and also on public records of birthrate, death rate, and total population. Information on the area planted to individual crops and crop yields was also being recorded in India.

The atmospheric circulation began to shift as early as 1873 in central Asia. While the summer rains were generally good over India in 1874, they were weak in some areas. In 1875 the weakening of the southwest monsoon resulted in substantially less summer rain in the Madras Presidency on the southeast side of India. The presidency lies in the rainshadow of the western Ghats Mountains, and the driest area is centered on the district of Bellary in the central Deccan Plateau. This year the rainfall was short enough in this district so that famine existed by the fall of the year.

For India as a whole, the two years of 1876 and 1877 constituted the driest two-year period during the 35-year period from 1861 to 1895, averaging only 86 percent of the 35-year mean. In Hyderabad and the United Provinces, rainfall averaged 68 percent of normal for the two-year period, and in Bombay Province only 77 percent of normal. The percent of normal rainfall does not adequately describe the drought potential. Where rainfall averages less than 500 millimeters, a small drop in the total and poor distribution of the rainfall through the growing season can result in nearly total crop failure.

The drought became more severe as the monsoon system persisted in a weak fashion in the summer of 1876. There was reduced rainfall again over a large part of southern India. The whole of the state of Mysore, half of Madras Presidency, and a large part of the state of Hyderabad and the Bombay Presidency were affected. By August the Indian government was well aware that a major famine was in progress and was going to get worse. In the state of Mysore, where the drought was in its second year, human mortality was high. The incidence of cholera increased as the summer went along adding to the perils. Mortality reached 20 percent by the end of the year. In Bombay Presidency all nine districts of the Deccan were stricken, as were some districts in southern Mahratta. The summer grain crops averaged only 38 percent of normal over the affected districts but were as low as 3.6 and 5.2 percent in the districts of Sholapur and Kalagdi. Since rainfall was so low, there was insufficient feed for livestock, and wells, ponds, and streams dried up. The lack of feed and water resulted in extensive loss of animals.

The monsoons were weak again in 1877. In Madras Presidency the situation was at its worst in the early months of the year. In the state of Hyderabad the worst conditions were in July and August. The drought and famine spread to the northern part of the Indian subcontinent, including parts of the Punjab, the North Western Provinces, the province of Oudh, and the state of Kashmir. The average annual rainfall was estimated to be 987 millimeters over the North Western Provinces, but during the summer monsoon season of June to September 1877, the total was 306 millimeters, less than one-third of the average. The rainfall was so low that there was no measurable summer crop harvested in many districts. Crop losses in other political units were substantial but not total. The shortfall of foodgrains in the north was placed at 3.35 million metric tons. Following the poor summer harvest, dysentery, smallpox, and cholera flared up into epidemic proportions taking a heavy toll of the population. In Bombay Presidency the excess mortality from the famine of 1876–1877 reached 800,000 lives over the normal death rate.

From November 1877 to December 1878 an estimated $1\frac{1}{4}$ million persons perished. In the first five months of 1878 there was a food shortage in nearly half of the districts of the Punjab. In

Madras Presidency, the rains increased in 1878 but remained below normal. The Famine Commission estimated that by the end of 1878 the loss of life in the Presidency totaled $3\frac{1}{2}$ million. Severe drought and famine continued in Kashmir in 1878 and famine conditions persisted into 1879 in some parts of India.

## Impact of the Drought in China

The change in the monsoons affected much of the rest of Asia. Part of China suffered drought and famine on a colossal scale. The provinces most affected—Shantung, Chihli, Honan, Shensi, and Shansi—cover an area of 777,000 square kilometers, an area nearly the size of France, and in 1875 contained a population of 100 million people. This region, often called the "Garden of China," receives some 70 percent of the total annual precipitation of less than 635 mm in the three summer months of June, July, and August. The remainder of the precipitation comes in the form of winter snowfall or spring rains in April and May. The actual distribution of precipitation varies through the year. In some years there is little snow, in other years there may be no rainfall in the spring, and in some years there is little rain in the summer growing season. The annual precipitation may vary as much as 30 percent from one year to the next.

Wheat, millet, and grain sorghum (kaolin) were the primary cereals in the north of China until after the thirteenth century, when maize, sweet potatoes, and peanuts were introduced. In the northern provinces the agricultural system was dependent on a winter wheat crop, planted in late fall and harvested in June, and millet and kaolin planted in April and harvested in September. If there was too little winter and early spring precipitation, there was a poor wheat crop, and if the summer rains were scarce, the millet and grain sorghum did not mature. In most years the agricultural production was just enough to feed the population. The failure of a single crop, either winter or summer, produced food shortages for both the human and animal populations. The food supply was marginal in the best of conditions, so there was never much grain in reserve.

In 1874 and 1875 precipitation had been below normal. In the fall of 1875 and spring of 1876 strong dry winds swept the plains of Chihli and Shantung provinces, destroying the winter wheat crop. With the drought extending into the summer of 1876 there was widespread failure of the crops throughout Shantung. Shantung is one of the largest of the North China provinces and by this time the population was nearly 27 million. In part of the province between 2 and 3 million persons suffered from malnutrition and at least $\frac{1}{2}$ million perished. By the fall of 1876 the drought-affected area had spread from Shantung to the provinces of Chihli, Honan, Shensi, and Shansi.

Problems became extremely severe in Shansi. Five to six million people were short of food. Typhoid fever became epidemic, adding to the distress of famine. By the fall of 1877 conditions were at their worst in China. The summer crops did not mature in much of Chihli and Honan. After the monsoons returned to more optimum conditions the governor of Shansi estimated that 80 percent of the population had experienced some hardship, and that 60 to 70 percent of the populace had contracted typhoid fever.

The failure of the monsoons through the years from 1876 to 1879 resulted in an unusually severe drought over much of Asia. The impact of the drought on the agricultural society of the time was immense. So far as is known, the famine that ravished the region is the worst ever to afflict the human species. The death toll cannot be ascertained, but certainly it exceeded 20 million.

Drought-related famine persists to the present day despite the technological advances of the Western world. The monsoon region of the Earth (Figure 8.1) includes a large part of sub-Saharan Africa. The population in this region is still largely dependent on subsistence agriculture for a livelihood. When drought occurs it means hunger, disease, emaciation, starvation, and even death. On a larger scale it may bring conflict within and between cultural groups, mass migrations of people and animals, and famine.

## DROUGHT IN THE SAHEL OF AFRICA

The Sahel of Africa (Figure 8.2) is one area where drought problems have become severe. The region was struck by a severe drought and extensive famine in the period 1968–1974 and again in 1983–1985. *Sahel* is an Arabic word which in this instance means "desert edge." The African Sahel is a vast area of plains and hills which stretches from the Atlantic Ocean on the west nearly to the Indian Ocean on the east. It consists of a variety of ecosystems, ranging from scattered clump grass to mixed trees and tall grass.

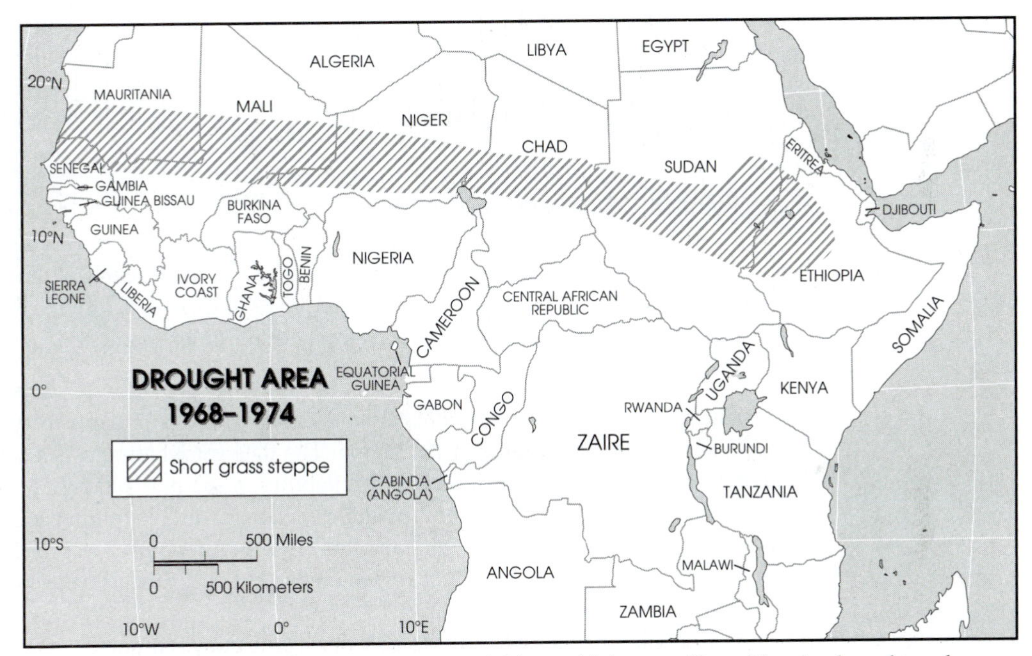


**Figure 8.2** The Sahel of sub-Saharan Africa, which was affected by the drought and famine of 1968–1974.

### Climate of the Sahel

The Sahel has a seasonal rainfall regime. On the desert margins there is only enough precipitation for the development of sparse grasses and the occasional acacia tree. The rainy season increases in length toward the equator. As the length of the rainy season and amount of rainfall increase, conditions become favorable enough for dryland farming.

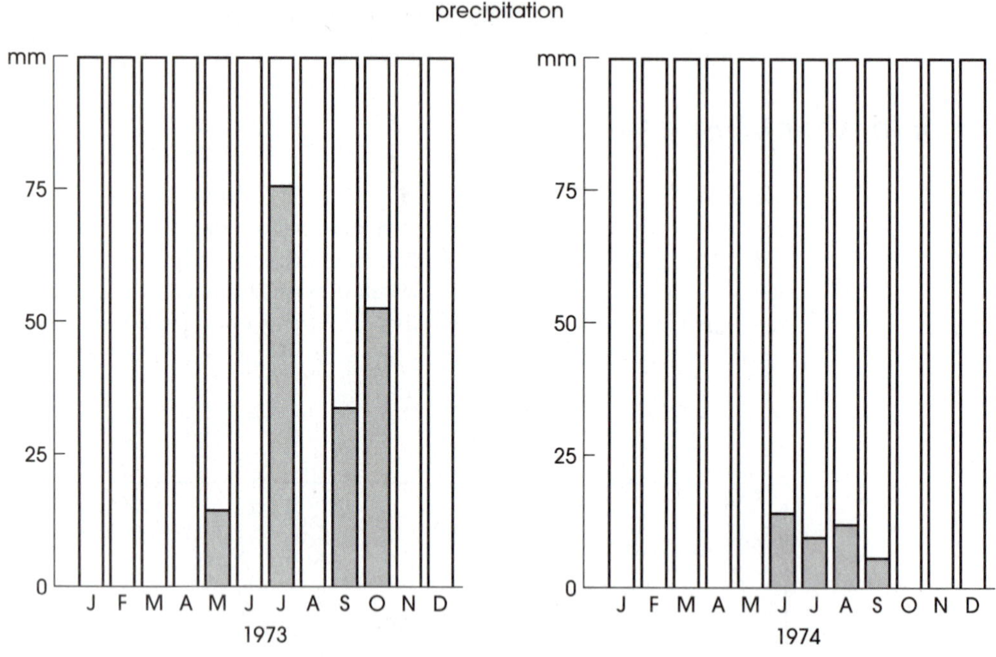


**Figure 8.3** Summer rainfall in Khartoum, Sudan in 1973 and 1974. Neither year was a good year for agriculture, although the rainfall seasons were very different. Rainfall was above average in 1973, but no rain fell in the months of June and August, which are critical months for agriculture. In 1974 there was rain in each of the four summer months, but the total was insufficient for crops to mature.

The precipitation is largely convectional and therefore of high intensity and short duration. It is also extremely scattered over the region. The data for Khartoum, Sudan (Figure 8.3) illustrate both the sporadic character and poor reliability of the precipitation. Most of the rainfall is concentrated in May through September, and while there was 175 millimeters of rain for the 1973 season, none fell in the critical growing months of June and August, so crops failed. In 1974, rainfall occurred in each month from June to September, but only 50 millimeters total was received—again a disaster.

Not only is the rainfall highly variable during the rainy season, but the total amount received each year is subject to marked fluctuation from year to year. The data for Khartoum again illustrate the problem (Figure 8.4). In 1973 the precipitation was three and one half times that of 1974. The nature of the precipitation is such that climatic drought occurs in some parts of the Sahel nearly every year. There may be ample precipitation near one village, and at a village only a few kilometers away the soil may be bone dry and the grasses withered. The climate is such that the chance that there will be enough rain in any given year to produce good crops and livestock forage is less than 50 percent.

The rainfall is also subject to cyclical fluctuations of varying lengths and intensity. Weather records indicate that dry and wet periods of several consecutive years were recorded during the years 1900–1903, 1907–1915, and 1930–1931. Precipitation was at least as low during these periods as during the years 1968–1974. Other periods of reduced precipitation have also occurred in the last 100 years.

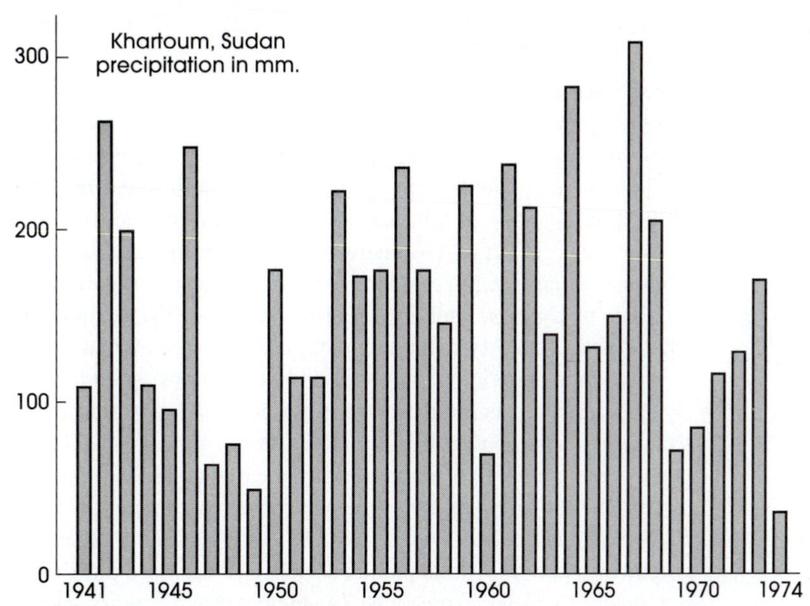


**Figure 8.4** Annual rainfall in Khartoum, Sudan, 1941–1974. The variability of precipitation from year to year is typical of the Sahel.

### Human Adaptation to the Drought

The size of the impact of the drought of 1968–1974 cannot be explained by the vagaries of the climate alone. There were other factors, primarily socioeconomic. The inhabitants of the grasslands are mostly either pastoral or agriculturalists. The pastoral people follow a seasonal grazing pattern governed by available pasture and water for the herds. Nomads have adapted to the seasonal pattern of precipitation and the scattered distribution of the rains. They design their entire migratory system to locate the patches of green vegetation that result from each rainstorm and to move from wet areas into the desert and back as pasture becomes available. Animals can survive in most ecosystems where there is at least a short season of rain. When an unusually dry year occurs, nomads move farther than normal in search of pastures or they can sustain themselves from the animals for a considerable period. The animal herds vary drastically in size, depending on available food and water.

At the onset of the drought in 1968 there were some 300 tribal groups totaling 20 to 25 million in the six West African countries. There were 17 million in Sudan and another 24 million in Ethiopia. The primary economic activity of 80 to 90 percent of the people was subsistence agriculture. Nearly 90 percent of the adult population was illiterate. Mali, Upper Volta, Chad, and Niger were among the 12 poorest countries in the world, with a per capita gross national product of less than $100.

In 1969 there was again an almost total loss of crops in parts of the Sahel, and reduced rainfall was observed over a larger area than during the previous year. By the end of the summer growing season in 1970 there were more than 3 million people needing emergency food rations in the western Sahel countries. After the failure of the summer rains of 1972, migrations became massive, producing social and economic problems. Disease and malnutrition became widespread. The total number of deaths will never be known, but all estimates place it in the tens of thousands and it may well have exceeded 100,000 in West Africa and another 100,000 in East Africa.

## DROUGHT IN THE 1980s

The specter of drought appeared again in Africa in the spring of 1980. This time the area affected was the Horn of Africa. The provinces of Harrar, Gamo, Goffa, and Sidamo in Ethiopia were the worst hit. In the earlier drought of 1973–1974, an estimated 200,000 people died of famine in Wollo Province. The drought claimed 80,000 cattle and affected some 1 million people. In parts of the adjacent state of Djibouti, rain did not fall in over two years. The starved livestock were the sole means of support for many of the people. In Somalia, an estimated 600,000 refugees lived in camps where the government provided at least a minimum of food and water. In Karamoja Province of Uganda, another 400,000 people faced starvation because of drought and civil war, which left the country without any organized governmental structure.

As in the previous drought of 1968–1974, rains failed for a succession of years in the 1980s. By 1985 famine was rampant on the African continent. Based on the total number of people affected, it was the worst disaster in the history of the African continent. Although climatic conditions were no more severe than before, the impact of the drought was worse. One of the primary reasons the drought was more severe was that the population of the affected area had grown from 60 to 90 percent in the 15 years since 1970. By March of 1985 perhaps 35 million people were persistently short of food, and human deaths had surpassed the 1 million mark. Not only was famine severe but it was widespread. Not a single African country had grain surpluses in 1984. Twenty-seven countries needed food aid.

As always during famine, people began to migrate. In Chad thousands died. Families abandoned infants and elderly as there was no food for them. Hundreds of thousands of people were wandering over the country looking for food. In Mauritania, the capital of Nouakchott grew in size from 100,000 to 400,000 between 1982 and 1984. Three hundred thousand people had come in off the grasslands looking for help. Thus a sixth of the entire population of the country was living in temporary camps around the city. In the spring of 1985 refugees from the famine were crossing the border from Ethiopia to Sudan at a rate of about 3000 per day. There were some 1.5 million Ethiopians in Sudan in relief camps or trying to get into them. In addition to the Ethiopians there were refugees from Uganda, Chad, and Zaire.

## SUMMARY

Drought is a problem in all areas of seasonal rainfall. Parts of Africa and Asia are particularly prone to drought of varying length and intensity. As time goes on and world population grows, the number of people affected by any change in weather or climate increases. The number of people depending directly on agriculture for a livelihood is also increasing. The impact of any fluctuation in rainfall in the semiarid lands thus becomes greater with time. While droughts of a few years duration do not represent a long-term change in climate, they certainly show what the effects would be of any major shift in rainfall patterns.

# part II

# THE HUMAN ELEMENT IN GLOBAL CHANGE

The relationship between the human species and the environment has been undergoing continuous change ever since the species appeared on the planet. As we know from Part I, the global environment has always been undergoing change. Sometimes the changes are sudden and catastrophic. At other times they take place over decades or centuries. These climatic and geologic changes have affected the human species since they first appeared a few million years ago. As the human population has grown, ever-greater numbers of people have been affected by any change.

The Pleistocene ice age influenced migrations of people. Some migrations were probably voluntary and others were forced by the ice sheets. Even the melting of the ice must have had major impact on humans as sea level rose, flooding hunting grounds and cutting off access from some land areas.

In the past few centuries droughts have claimed millions of human lives. Unusually cold weather created great hardship at other times. Today, any global change affects tens of millions of people, if not the entire species.

During the early part of human history natural changes through mutation allowed the species to begin to exert some control over its habitat. In the last portion of human history, technology rather than physical adaptation has been the instrument used to modify our environment.

The study of the origins of certain basic inventions underscores the fact that historical change is a continuous process, at once dynamic and often unpredictable. Technology in the form of massive irrigation schemes transformed parts of the world's deserts into some of the richest agricultural regions of Earth. Development of fast-growing and rust-resistant varieties of grain has permitted the extension of the grain belt of North America much farther north than in previous decades.

Technology has created artificial environments. These artificial environments can travel in space between Earth and the moon, beneath the surface of the world ocean, and in arctic conditions without essential changes in human physiology.

A great deal of technology is directed at changing the natural environment. Humans work to reshape the surface of the land, change the courses of rivers, and alter the living flora and fauna. The assumption is always that we can improve on nature, that it is not suitable in its own way. The rate at which we have altered the environment has paralleled the growth of the human population. We are now creating more different kinds of changes in more places than ever before.

Some changes taking place now are natural. Others may be human induced. Some may be natural changes enhanced by human activity. The impact of these changes becomes greater each day in terms of economic and social costs.

chapter 9

# Industrial Revolution and World Population Growth

***D. Gordon Bennett***

***CHAPTER SUMMARY***



Once the agricultural revolution got under way, the rate of technological development began to increase steadily. For instance, waterwheels were the first form of inanimate energy used to increase production and to process raw materials. Waterwheels for powering grain mills existed by the time of Christ. Windmills used for the same purposes came into existence in the tenth century. Both waterwheels and windmills powered a variety of machinery.

One of the major results of technology is communication. Increased mobility started an unprecedented period of exploration that eventually carried human beings to the moon and back safely. For several hundreds of years, inventors attempted to make an accurate clock. Finally, in the fifteenth century the escapement mechanism was invented. This device was the basis for accurate chronometers, which allowed sailing vessels to go far from the sight of land and still determine their location. In 1492 Columbus made the historic voyage to the western hemisphere. Shortly after, in 1497, Vasco de Gama sailed around the tip of South Africa.

## FACTORS CONTRIBUTING TO WORLD POPULATION GROWTH

Around 1650 the human population reached the 500 million mark (Figure 9.1). It had taken 3 to 4 million years for the species to reach this size. Largely in response to increased agricultural production, the rate of population growth was increasing rapidly. The industrial revolution proper began in the period 1783–1812 in what is now Great Britain. It began slowly and grew faster and faster. By 1830 the world population passed the 1 billion mark. The industrial revolution resulted in a combination of factors that significantly lengthened the average life span and increased the rate of population growth.

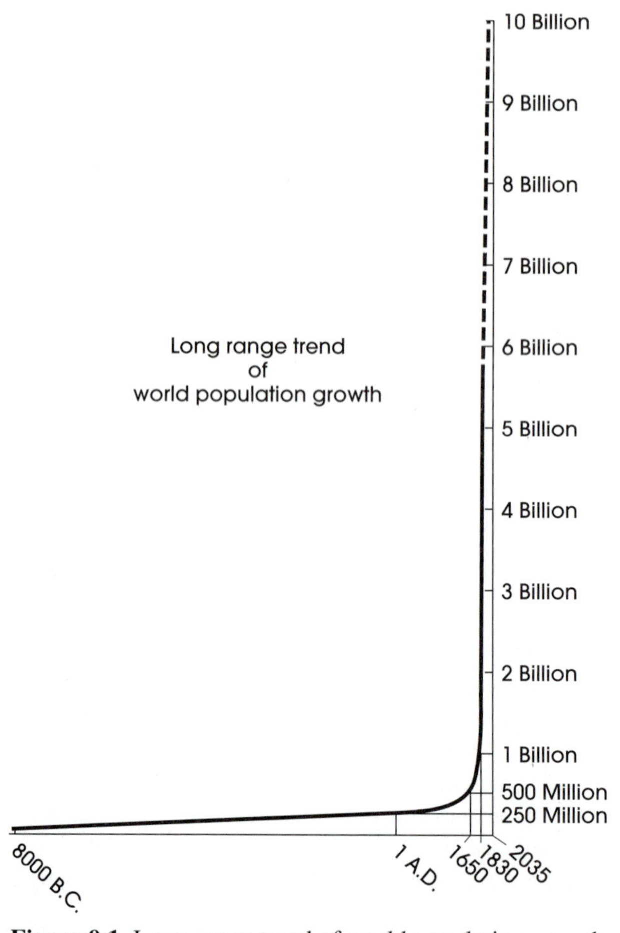


**Figure 9.1** Long-range trend of world population growth.

## Emigration and European Population Growth

The newly discovered lands of Africa, the Americas, and Australia provided a place to send the surplus population of Europe. The new lands experienced rapid population growth. This migration produced an almost instantaneous population burst into some of these colonial areas. The classic example of this rapid settlement were the land runs in Oklahoma. Individuals and families claimed thousands of square kilometers of land in one day. Guthry and Oklahoma City, Oklahoma, appeared in the space of a few hours. This certainly was a phenomenon unique in world history.

## Technology and Industrial Production

As population grew and new lands needed to be settled, the world demanded more and better machinery to increase production and to further increase mobility. People needed to move more goods faster. Laborsaving machinery to produce agricultural commodities was also needed. The industrial revolution got into full swing in the Western world with the invention of the steam engine by James Watt. Steam moved goods over the land, over the seas, and did a variety of other tasks. James Watt patented a steam plow as early as 1784. In 1794, Eli Whitney invented the cotton gin. American cotton production rose from some 2.27 million kilograms in 1794 to 100 million kilograms in 1825.

Cyrus McCormick invented the reaper (the predecessor of the modern combine) in 1831 and it went into commercial production in Chicago in 1845. By 1870 steam tractors came into use and in 1907 Ford tractors had gasoline engines. Agricultural technology has cut the time needed to plow 1 hectare of land from 14 hours to less than 1 hour.

## Implementation of Public Health Measures

The effects of technology on birth and death rates can be sudden and astounding. The development of immunization is one major advance. Immunization against communicable disease was first realized in 1796 when Edward Jenner showed the use of inoculation. Pasteur further developed the process in the 1880s, but it did not become widely used until well into this century. For centuries, typhus, or spotted fever, took a large toll of lives. This was particularly true among soldiers and refugees in time of war. The disease was particularly widespread from the fifteeth century until World War I. The disease struck the Serbian Army during the war. The death rate in some places reached as high as 75 percent. Charles Nicolle discovered that body lice (*Pediculum humanus corpora*) carried the disease. Improved standards of cleanliness helped eradicate the disease.

The development of DDT as a pesticide to control mosquitos and other insects provides another example. Before the introduction of DDT on the island of Mauritius, there was slow population growth. The birthrate was 33 per 1000 and the death rate 27 per 1000. In the three years from 1946 through 1948 the government treated the island with massive doses of DDT to kill the *Anopheles* mosquito. By 1952 the birthrate had risen to 47 per 1000 and the death rate had dropped to 12 per 1000. In 1965 the death rate had gone down still further, to 8.6 per 1000. During the seven-year period the annual growth rate of the population had rocketed from 0.5 percent to nearly 3 percent. This was due largely to the reduction of infant mortality. This dropped to nearly half what it had been before the introduction of mosquito control measures. The drop in the death rate in the seven years from 1945 to 1952 equaled that which England and Wales took 100 years to accomplish. Mauritius is not the only location where such drastic drops in the death rate took place.

They began a mosquito eradication program in Ceylon (Sri Lanka) in 1946. In two years the death rate dropped from 22 per 1000 to 14 per 1000. The population doubling time dropped from nearly 150 years to 23 years.

## URBANIZATION

In 1800, nearly half of the world's land surface was still unknown to Europeans. In the next century more land was explored and colonized than in the previous 350 years. The nineteenth century began a period of migration that would result in the largest redistribution of people in human history. Millions of Europeans left their homes and crossed the oceans to settle in these new lands. During most of human history people were rural dwellers and dependent on the land for a livelihood. In 1800, only about 3 percent of the world's people lived in urban areas. Less than 2 percent lived in cities of 100,000 or more. Urbanization in the Western world grew along with industrial activity. Industry provided jobs that attracted people away from rural areas, and industry needed inexpensive labor. In the cities, jobs were available but living conditions were very poor and death rates were high. The high mortality resulted from a poor diet, lack of sanitation, and crowding. These conditions aided in the spread of contagious diseases. The death rate in cities was so high that in-migration accounted for 90 percent of urban growth until as late as 1850.

The percentage of world population living in cities quadrupled in just 100 years. By 1900, 14 percent of the world's population was living in urban areas of 5000 or more. As the flight to urban areas continued, some cities grew to large size. In 1950, there were 71 cities with populations of 1 million or more. Two cities, New York and London, reached 10 million and were the major cities of the industrial world.

## THE DEMOGRAPHIC TRANSITION

As societies change from agrarian to industrial, there is a shift in population growth rates that have a well-defined pattern. The model of this shift shows a change from high birth and death rates to low birth and death rates (Figure 9.2). This change takes place in several stages, known as the demographic transition. The model of the demographic transition is based on events in Europe. In the first stage, birthrates and death rates are high and fluctuate widely. This was the situation in most of the world before 1750. The total world population was growing slowly but erratically.

During the second stage, death rates fall rapidly while the birthrate remains high. In much of Western Europe, the death rate dropped sharply after 1750 due to advances in medicine and public health, but the birthrate stayed high. This greatly increased the rate of growth, from about 0.1 percent per year in 1750 to about 1.4 percent by the third quarter of the 1800s. The growth rate in these countries exploded by some 14 times. European colonies were also experiencing rapid increases. As emigration from Europe increased and there were further gains in agricultural and industrial productivity and health care, the death rate continued to drop in these countries. During this stage, the rate of population growth increases and total population increases geometrically. The growth rate in the European countries and their colonies was far greater than in the rest of the world at this time.

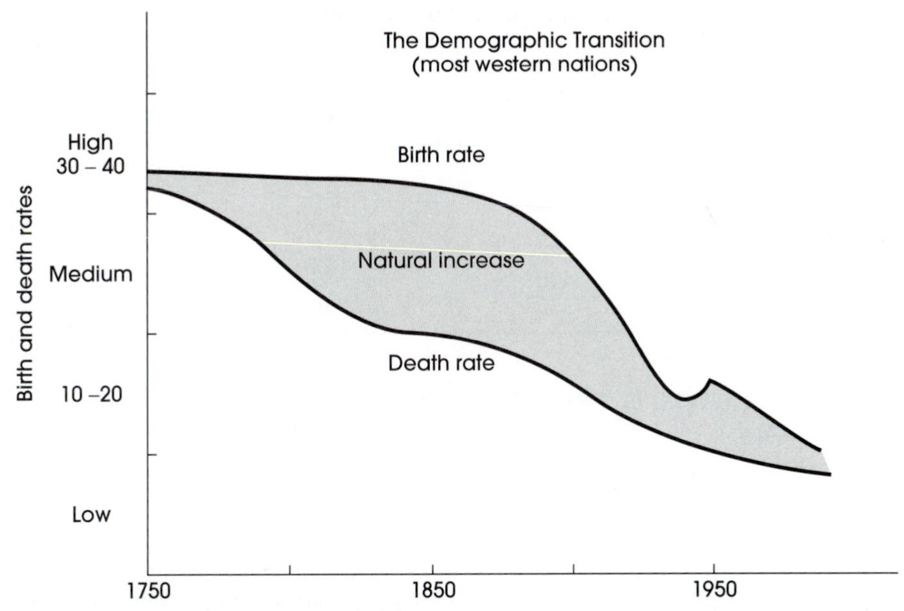


**Figure 9.2** Stages of the demographic transition in many industrialized countries. In stage A, both birth and death rates are high. In stage B, birthrates are high and the death rate drops, resulting in a rapid population expansion. In stage C, both birth and death rates are low.

The third stage of growth occurs when the birthrate drops while the death rate continues to decline, but more slowly than before. By the last quarter of the 1800s, the birthrate in Europe began to decline, thus slowing the growth rate. The exact reasons for the drop in the birthrate are not known, but several factors are important contributors. Among these are:

1. The more crowded living conditions in the growing industrial cities compared to the rural countryside
2. The greater expense of rearing children because of the new child labor and compulsory education laws
3. The realization that it was not necessary to bear as many children as before in order to have a son live long enough to care for his parents in their old age

In the final stage the birth and death rates are low, with the birthrate slightly higher than the death rate. Once again the difference between the birthrate and the death rate is small. The drop in death rates resulted in the average human life span increasing from about 30 to more than 70. By 1930 the industrialized countries of Europe had passed through the transition and reached the final stage of low birth, death, and growth rates. Those countries that have joined the industrial nations since 1930 have also gone through a similar demographic transition. Since 1900, growth rates in the industrialized countries have averaged between 0.25 and 1.0 percent.

By 1930, world population had doubled again to 2 billion, cutting the time required to double the population nearly in half. In addition, twice as many people had been added during the preceding 100 years as in the prior 180 years (Table 9.1).

**TABLE 9.1** World Population Growth, A.D.1–1994

| Date | Population (millions) | Years to double since last date | Average rate increase per year (percent) | Number increase since last date (millions) |
|---|---|---|---|---|
| A.D.1 | 250 | — | — | — |
| 1650 | 500 | 1650 | 0.04 | 250 |
| 1830 | 1000 | 180 | 0.4 | 500 |
| 1930 | 2000 | 100 | 0.7 | 1000 |
| 1976 | 4000 | 46 | 1.5 | 2000 |
| 2021[a] | 8000 | 45 | 1.6 | 4000 |

[a] Projected.

## POPULATION GROWTH IN THE UNITED STATES

In the United States, population changes followed a similar pattern as in other industrial countries. Around 1820, the total fertility rate (the average number of children per woman) was nearly 7. Death rates at that time were very high, nearly offsetting the high fertility rate. Fertility began to drop after 1820 and by 1900 women were having an average of $3\frac{1}{2}$ children. During the depression years of the 1930s the average dropped to just slightly more than two children.

Fertility increased slightly during World War II, then rapidly during the baby boom years of 1946–1957. The rate went back up to nearly three children in 1947, and then to more than $3\frac{1}{2}$ children in the peak year of 1957. This was the highest level since the late nineteenth century. Several factors led to, and sustained, this period of rapid growth. It began with the return of many military personnel from war zones overseas. The years after World War II were years of prosperity, and families could afford to raise more children than during the long depression of the 1930s. More important was the change in outlook for the future. Young adults reared during the depression and world war now enjoyed a time of peace and rising real incomes.

Fertility dropped sharply between 1957 and 1976 (to under two children per female) and then stabilized. Births in the United States still exceed deaths by 1.7 million each year. Including a small rise in fertility since the mid-1980s and a major increase in legal immigration since the 1950s, population growth in this country is now more than 2.8 million each year. When a conservative estimate of illegal immigration is included, the annual net increase in the number of people living in the United States is more than 3 million.

The process of urbanization took place quite rapidly in the United States as it did in other industrial countries. In 1790, more than 90 percent of the population lived on farms. Today, the ratio is less than 4 percent. Not only has the relative number of people living on farms decreased but so has the absolute number.

## POPULATION GROWTH IN THE DEVELOPING COUNTRIES

After World War II ended, most European colonies sought and gained independence. In the 15 years from 1945 to 1970 the number of independent countries increased rapidly. The number of independent nations has doubled from about 90 to 180. Much of the technology in health care gained during the war was available to the developing countries. This included such basic health care innovations as the use of pesticides, inoculations, and basic sanitation measures.

Many of the developing countries entered the second phase of the demographic transition after World War II. They experienced a more rapid decline in the death rate than that experienced in Western Europe. The result was the developing nations containing most of the Earth's people exploded in size. The developing countries contained about two-thirds of the world population in 1940. By 1990 this increased to about three-fourths of the total.

Within only 46 years, another doubling of world population occurred. This was less than half the time required for the previous doubling and only about a fourth of the time necessary for the one before that. From 1930 to 1976, the annual growth rate for the world population was 1.5 percent. It was 0.7 percent from 1850 to 1930. Moreover, the number of people added to the planet during the 46-year period from 1930 to 1976 was 2 billion. This was twice the 1 billion added during the previous 100 years and four times the 0.5 billion added in the previous 180 years (Table 9.1). Thus not only was the doubling time decreasing and the rate of growth increasing, but the numerical increment of the doubling was also becoming gigantic.

In the early 1970s, the world population growth rate peaked at about 2 percent annually. During the 1970s, family planning was accepted in some of the developing countries and growth rates began to decline. Among these countries were the People's Republic of China, Sri Lanka, South Korea, and Taiwan. Within less than a decade, the world growth rate dropped to 1.8 percent, where it remained until about 1990. It has since dropped to 1.6 percent. Although the growth rate has declined in the past few years the doubling time is still very short. It is currently about 45 years. If current rates of growth continue, world population will double from the 1976 total of

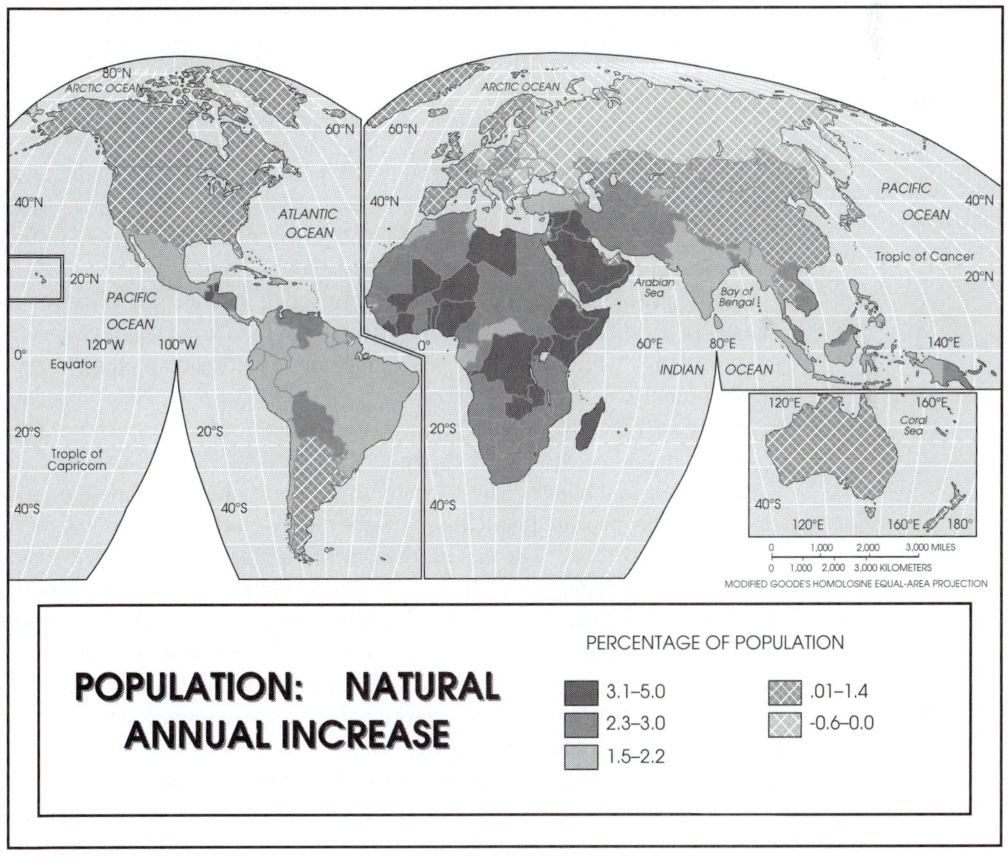


Figure 9.3

4 billion to 8 billion by 2022. In 1994 the World Bank forecast world population to reach 8,474,000,000 in the year 2030.

There is considerable variation in birthrates at present. The highest birthrates occur mainly in Africa and southwestern Asia. The highest death rates occur in many of the same countries. Growth rates in Africa and southwestern Asia are as high or higher than those reached at any time by European countries during their demographic transition (Figure 9.3). Several countries in Latin America are also experiencing high rates of increase. Birth and death rates are lower than in Africa, but the gap between the rates is just as large.

Growth rates do not tell the complete story. The largest numerical growth in population does not occur in the countries with the highest growth rates. Total world population is growing at about 90 million per year. Even though the growth rates are not as high as in some other countries, more than a third of the total increase in world population each year occurs in two countries, India and China. This is because these countries have such a large population. A relatively small growth rate produces very large increases in absolute numbers.

In the industrialized countries, population growth rates have dropped enough so that many national populations have nearly stabilized. As a result, over the past two decades the United Nations has lowered its estimates of world population for the year 2000 from 7.5 billion to 6.1 billion. In 1990 the province of Quebec, Canada began offering cash bonuses for live births. In Singapore the number of children per family dropped from 4.7 to 1.4 in less than 30 years. There is now a government-operated match-making service intended to increase the birthrate of the native population.

## *THE POVERTY/HUNGER GAP*

Nations and regions have gone through the industrial revolution at different times in history and at different rates. Different regions of the Earth have differing resources to support their local populations. The result of these different development rates and different resource bases is that there are large disparities in wealth among the human population.

The people of the world have been living in increasingly different societies: one rich and one poor. The difference is primarily between the industrialized developed nations and the agrarian developing countries. There are strong contrasts among nations in per capita gross national product (GNP), a measure of economic wealth (Figure 9.4). Nearly half the world's people live in countries where the per capita GNP is less than $500 a year. About 10 percent of the population lives in nations with a per capita GNP more than 40 times that much! Part of the reason for the economic and hunger gap is the continued high rate of population growth in the developing countries. The growing population absorbs most of the added national production.

Western Europe sent tens of millions of its citizens abroad as it passed through the expansive stage of its demographic transition. This is not an option for the rapidly growing developing countries today because no unclaimed empty areas exist to which to send their surplus populations. This puts more pressure on food resources, industrial development, and employment opportunites in these countries. Unlike in the eighteenth century, the peoples of Earth no longer have access to vast, rich, empty lands to which to go to raise more food. There are no unoccupied continents which they can exploit to import the additional resources they need. Nor are there empty lands to which they can export their surplus human population.

Most nations of Western Europe could relieve the pressure on the land by making new industrial jobs available in the cities to which the surplus rural population moved. Developing countries have not been able to expand urban job opportunities rapidly enough to meet the demand

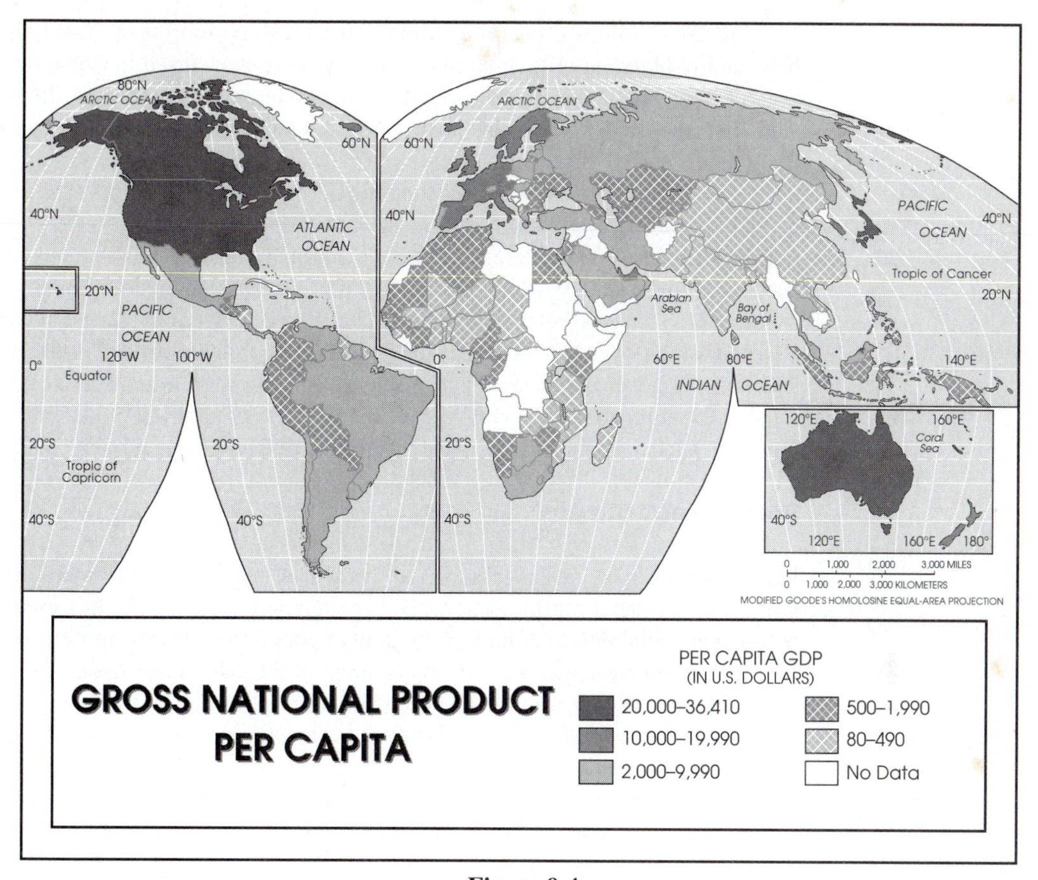


**Figure 9.4**

caused by the influx of the rural masses from the land. Unemployment combined with underemployment averages 40 to 60 percent in most of these poor countries, and during the next 30 years, the number of people entering the labor force will double.

However, there are disparities within developing countries as large as those between the developed and developing countries. Indeed, the gap is greatest within the poorest countries, where leaders have often been those who have most strongly criticized the international disparities of income and of consumption. Moreover, within these poor countries, the per capita income of the wealthiest 20 percent of the population is frequently 20 to 25 times that of the poorest 20 percent. This is especially true of the nations of Latin America, Africa, and southwestern Asia. Within the developing countries the nature of the economic gap is between the urban industrial population and the rural agricultural population.

### Growth Rates in the Developing Countries

There is some hopeful optimism for the reduction of growth rates in the developing countries. World growth rates have dropped some 25 percent from the peak a few decades ago to the present. The growth rates of the poorer nations are still three to five times those of the richer countries. Thus birthrates will have to continue to fall faster than death rates if there are to be further reductions in the growth rates of the developing countries.

Today, developing countries containing over three-fourths of those living in the poor lands have family planning programs. Major successes have occurred in Columbia, Thailand, Taiwan, South Korea, Sri Lanka, and several other small countries, as well as in the People's Republic of China. More recently, somewhat lower birthrates have also been noted in Egypt, India, Brazil, the Philippines, and several other countries.

One major setback to family planning occurred from the mid-1980s to the early 1990s under the Reagan and Bush administrations, when family planning funds were denied to family planning organizations in this country. Family planning funds were also denied other countries which used *any* of their money—whether from the U.S. government or not—for any abortion-related activities. Many experts feel that this resulted in fewer people having access to family planning methods than would otherwise have been the case. This leads to an increase in births, larger populations, and more hunger and environmental problems. In 1994, the Clinton administration once again permitted U.S. aid funds to be given to the United Nations Fund for Population Activities and to similar agencies that are trying to reduce the growth rate in the developing countries.

## Present Food Supply

Per capita calorie supply is less than 90 percent of requirements in more than 25 countries that contain more than a fourth of the world's population (Figure 9.5). It is inadequate in more than 60 nations containing over half of the human population. There are rich countries in which the average person consumes 120 percent or more of the calories necessary for an adequate diet.

The industrial revolution has been largely responsible for the exponential growth in population and in production. It created many new resources that increased our quality of life. Technology, however, does not provide answers to the basic problems of the human species and the environment. Technological answers to environmental problems often only postpone the inevitable or generate problems as bad or worse than the original problem.

Just a few years ago, high-yielding varieties of wheat and rice were developed which were supposedly going to start a "green revolution." Some agricultural experts predicted huge food surpluses for the world. Coming at a time when per capita food production in the developing world was declining, the Green Revolution was heralded as an exciting advance. Using the new hybrids, India, a country that suffered chronic food shortages, doubled its wheat production in a six-year period. On the basis of these high-yielding cereals and a few years of above-normal rainfall, India announced that the country had become self-sufficient in the production of food. Other countries increased cereal production almost as dramatically.

These high-yielding varieties of grain need a lot of fertilizer and water. It takes energy to produce artificial fertilizers and the usual energy source is natural gas or fuel oil. Beginning in 1973 with the oil embargo, some fossil fuels became scarce, which reduced the supply of fertilizers. What was more important than the reduced supply was the rapid increase in price. In the United States, nitrogen fertilizers doubled in price between the summers of 1973 and 1974. In the developing countries price increases were even greater. In some countries the price tripled. Norman E. Borlag, 1970 Nobel Peace Prize winner for his work in developing the high-yielding grains, estimated that in 1972 it cost India almost $600 million to import crude oil, fertilizer, and grain. In 1973, after only a few months of mushrooming oil prices, the cost of the same imports was nearly double that of 1972. For 1974, costs nearly doubled again.

The impact of these price increases on farmers living at or near the subsistence level was devastating. The prices for agricultural staples did not rise at the same rate as the cost of production. In Sudan, for example, the government fixed the price of wheat to keep bread costs down. Farmers were squeezed between a rising cost of producing commercial crops and a fixed price

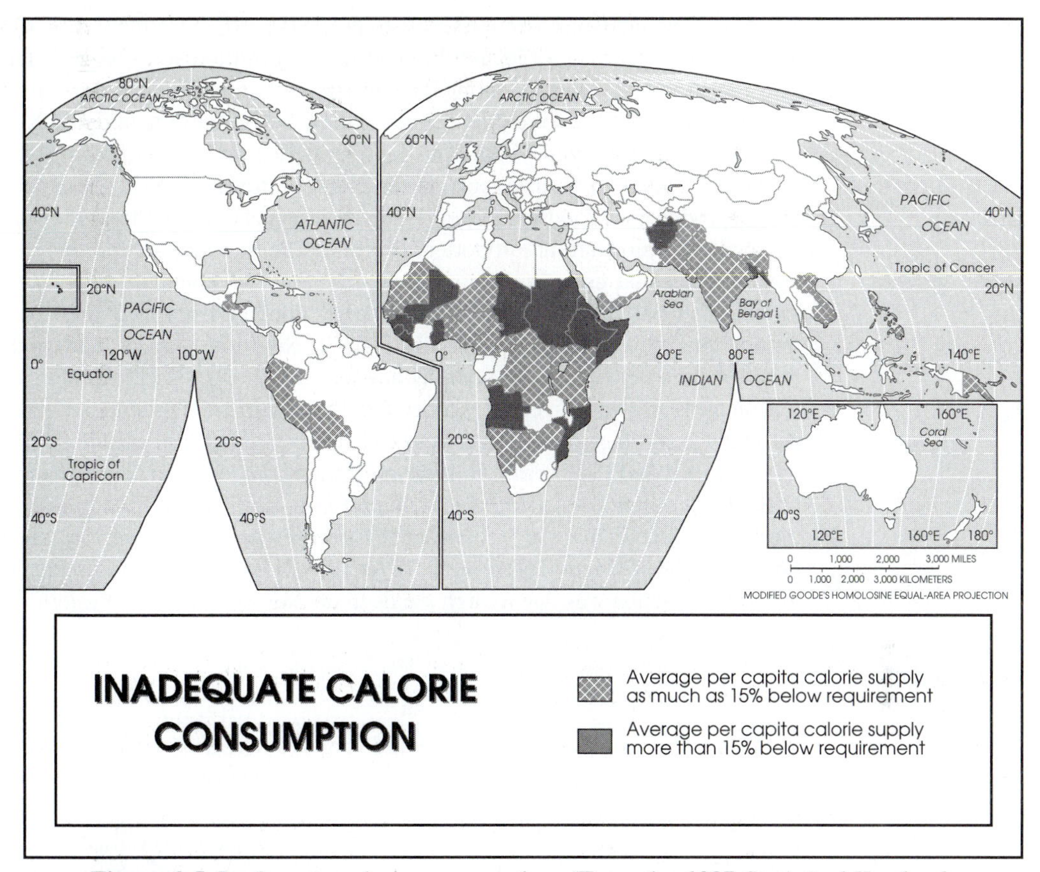


**Figure 9.5** Inadequate calorie consumption. (From the *1987 Statistical Yearbook*, United Nations, New York, 1990.)

for the crops. The results were predictable. They gave up producing commercial crops, quit buying fertilizer, and returned to subsistance crops. The end result was less total food production in the very countries that needed it most. These developments may have set tropical agriculture back many years because farmers took such a loss on the high-yielding grains during the oil price war.

In the developing countries, population was growing at about the same rate that agricultural productions was growing. As a result, per capita food production in 1991 was about 13 percent greater than it was in 1970. In contrast, although food production in the developed countries only grew by about 25 percent in the same period, a lower population growth rate resulted in a per capita food increase of about 10 percent. Today, there are still between 750 million and 1.3 billion people who are hungry. The net effect of the Green Revolution was to delay worldwide food shortages for a few decades.

Food shortages are no longer limited to the developing countries. The five-year period from 1970 to 1975 saw the world supply of grain in storage drop from 90 days to 35 days. In 1970, Canada reduced the amount of land planted to wheat by 4.9 million hectares, due to a glut in the world wheat market. Wheat was piled in huge mounds in the Great Plains, much of it slowly rotting. By 1973 the amount of land planted to wheat had nearly doubled. From August 1972 to August 1973 prices received by farmers for agricultural commodities jumped an average of 62 percent. Wheat escalated from $1.51 to $4.45. Soybeans increased from $3.36 to $8.99, corn from

$1.15 to $2.68, and eggs from 29 cents to 60 cents a dozen. In the five years from 1970 to 1975 idle grain equivalent of cropland dropped from 71 million acres to zero. This grain land represents an additional 21 days' supply that was brought into production over this five-year period. In 1994, world food reserves were again at the low levels of the mid-1970s.

Between 1970 and 1976, both Latin America and Eastern Europe changed from grain-exporting regions to grain-importing regions. The former Soviet Union and China, two of the largest producers of food in the world, are importers of grain. Of the 115 countries for which data are available, most are net importers of grain. The two major exporters of grain today are the United States and Canada. Grain exports by these two countries have increased rapidly in response to global demand. Grain exports nearly doubled from 1970 to 1976, increasing from 56 million tons to nearly 100 million tons in 1976. Since the United States and Canada provide so much of the surplus grain, a small decline in production in North America can have a major effect on the world trade in grain. A 5 percent decrease in grain production in North America would amount to a shortage of 20 percent of the grain available for international trade.

In 1975 and 1976 grain production was at record levels in North America. Corn production in the United States reached 5.77 billion bushels in 1975 and 5.89 billion bushels in 1976. Increased production involved record, or near-record, yields of wheat, rice, and sugar. The FAO reported that the production of oilseeds, vegetable oils, and coffee was also up. Much of the increase in food production was due to high yields in the Soviet Union and Eastern Europe.

## *SUMMARY*

In the last 150 years the human population has been growing at an ever-increasing rate. A number of factors have led to this rapid growth. The advent of the industrial revolution is a primary element. The industrial revolution led to greater production of food. It created jobs in the factories. The increasing population led to greater demand for industrial products. Public health measures reduced the death rate. The discovery of the western hemisphere lands provided a place for surplus population and actually created a demand for labor.

Today, the fastest-growing countries are the developing countries. Many of the nations with the highest growth rates are in Africa. China has the largest population of any country. India has a smaller population than China but a higher growth rate. In a few years it will be the most populous country.

The human population reached its highest annual growth rate, a little over 2 percent per year, in the late 1970s. Even though the percentage of growth has begun to decline slightly, the absolute number of humans added to the planet each year continues to increase.

## BIBLIOGRAPHY

Ehrlich, P R. and A. Ehrlich. 1990. *The Population Explosion.* New York: Simon and Schuster.

Peters, G.L., and R. P. Larkin. 1993. *Population Geography.* Dubuque, Iowa: Kendall/Hunt Publishing Co.

Population Reference Bureau. 1995. *World Population Data Sheet.* Washington, D.C.: Population Reference Bureau, Inc.

U.S. Department of Health and Human Services. 1990. *Healthy People 2000.* PHS-91-50212. Washington, D.C.: Public Health Service.

U.S. Census Bureau. 1989. *World Population Profile 1989.* Publication #003-024-07074-0. Washington, D.C.: Census Bureau

chapter 10

# Land Degradation: Accelerated Soil Erosion

***CHAPTER SUMMARY***



Natural soil is the product of many elements and processes. Soil forms slowly from chemical, physical, and biological processes. The rate at which it forms depends on the local conditions of climate, vegetation and topography. An average rate at which topsoil forms is from 10 to 20 millimeters each 1000 years. The world has hundreds of different types of soil, each with its own unique qualities. These unique qualities provide the differing elements that make up the soil profile. Worldwide, there is an average depth of 178 millimeters (7 inches) of soil over Earth's land surface. This totals some 400 metric tons per hectare (1110 tons per acre) over a total of 1.2 billion hectares (3.1 billion acres) of cropland. The maximum amount of soil with which to produce food and fiber is 3.2 trillion metric tons.

## *MECHANICS OF SOIL EROSION*

Soil erosion is the weathering removal of soil by wind and water. Wind and water move the eroded particles to some other location. There it is deposited as sediment. Soil erosion is a natural process that removes soil from the land surface. Natural rates of soil removal from land with a good cover of vegetation are less than those of soil formation. The natural process is for soil to accumulate slowly with time.

The primary means by which humans change the global environment is through the process of producing food for the growing population. Increased soil erosion is a product of growing food. Due to soil erosion the world supply of soil decreases each year.

### Erosion by Water

Erosion by water involves three major processes: removal, transportation, and deposition. The process of erosion begins whenever rain falls at a rate greater than the soil can absorb and water begins to flow over the surface. The process ends when the running water deposits the eroded material at some other location. The deposition may be a few centimeters away or thousands of kilometers away in the sea.

### Entrainment

Entrainment is the removal of particles by the friction of water over the particle. The amount of lift depends on the velocity of the water. The development of eddies aids the lifting process by increasing velocities. Fine sand and coarse silt are actually easiest to erode, requiring a critical turbulent velocity of around 300 millimeters per second (Figure 10.1). Fine silts and clay-sized

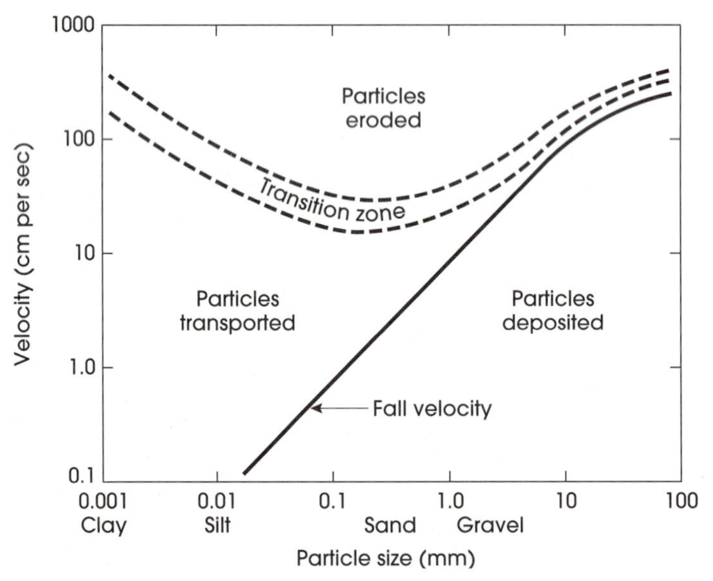


**Figure 10.1** Velocity of water required to erode soil particles and to keep them in suspension. It takes a higher velocity to erode clay and silt-sized particles than to erode sand or gravel. Smaller particles will stay in suspension at lower velocities than will sand or gravel.

particles are more difficult to erode because they are so small. They lie within the laminar or stagnant layer of water on the bottom, where there is no friction to lift the particles. Larger particles of sand and gravel are also difficult to erode. This is because of the greater mass of the particles.

## *EFFECTS OF HUMAN ACTIVITY ON SOIL EROSION*

Whenever and wherever water flows over cultivated land it causes erosion. Agriculture increases the rate of erosion by altering some of the variables that affect the erosion process. Soil is most susceptible to erosion following preparation for planting. Plowing, disking, and other tillage operations loosen and detach the soil.

### Sheet Erosion

Erosion from the land surface will occur even within sheetflow. Sheetflow is water flowing over the ground in an unchanneled sheet of water. Sheet erosion is the steady reduction in the level of the surface by sheetflow (Figure 10.2). Although the layer of moving water may be thin and velocities low, it entrains fine particles. This type of erosion is very active in drylands and also near

**Figure 10.2** Sheet erosion from a wheat field in Deaf Smith County, Texas. The foreground of the picture was a roadside ditch about 1 meter deep. A rainstorm the day prior to the day the photograph was taken produced 15 centimeters of rain. It washed enough topsoil off the field to fill the ditch completely with sediment.

melting snowdrifts in mountain regions. It is often difficult to detect on the surface until considerable damage is done. In some areas sheet erosion removed up to half a meter of soil.

### Rill Erosion

When sheetflow begins to collect in rivulets, slope channeling or rill erosion may develop. Rills are small channels up to several centimeters deep in the soil. It may proceed fast enough even during a single storm to require replowing. Plowing may overturn the soil to depths of up to 30 centimeters. Up to 7 metric tons of soil per hectare (20 tons per acre) may erode from farm fields during winter and spring runoff. This erosion may go unnoticed since sheet and rill erosion is not visible after plowing. Sheet and rill erosion remove nearly 3.6 billion metric tons (4 billion tons) of topsoil globally each year. If ignored, rills become gullies and running water will cut deep gashes in the land. About half the loss occurs on cropland. Rilling may proceed through the topsoil. Left without protection, rills develop into gullies that cut through the soil to bedrock. Badlands are the full development of gullying so that no flat surfaces remain.

## *FACTORS THAT AFFECT SOIL EROSION*

The main factors that affect the amount of surface erosion are:

1. Type and amount of vegetal cover
2. Amount and intensity of precipitation
3. Length and steepness of slope
4. Type of soil
5. Tillage practices used

The major change that agriculture brings to the land is the removal of vegetal cover. Actual measurements of erosion under different types of vegetation vary widely. One result of the removal of vegetation is to reduce the infiltration capacity of the soil. Precipitation intensity exceeds the infiltration capacity of most soils, with complete cover of vegetation only when there are extreme rainfall events. On bare soils rainfall reduces the infiltration capacity of the soil by sealing the openings in the surface with sediment.

The more precipitation there is, the more erosion there will be because of more water running off the surface. The more intense the precipitation, the more erosion. As intensity increases, the percentage of precipitation that runs off increases.

The longer a slope is, the more soil loss there is. If the length of slope doubles, soil loss increases about one and half times. The steepness of slope is, however, more important than length. Land with slopes of 0 to 3 percent is level or nearly level and suffers little soil loss because surface runoff moves very slowly. Land with slopes of 3 to 5 percent is gently undulating to rolling. On land with slopes in this class, sheet erosion is common and rill erosion may develop. On slopes as steep as 10 to 15 percent severe gullying may occur. Slopes in excess of 15 percent face nearly certain destruction of the soil by rapid erosion if vegetation is removed. If the steepness of the slope doubles, soil loss increases about 2.5 times.

Actual rates of erosion in the United States average 60 millimeters per 1000 years, two to three times higher than can be expected from natural processes. In small loess-covered basins in Iowa (basins of less than 3.4 square kilometers in size) erosion rates of 12.8 meters per 1000 years have been measured. Annual soil loss in the United States has been estimated at 1 billion tons. Of this total, half is carried by the Mississippi River. This represents a rate of 400 millimeters per 1000 years, a rate 20 times the natural rate from a forested watershed.

## MOVEMENT OF SEDIMENT

Of the total soil eroded from hillsides by water, 10 to 50 percent ends up in streams. Streams move soil particles in several different ways: (1) as dissolved mineral compounds, (2) as suspended matter, and (3) as bedload (saltation).

### Dissolved Load

The type of material transported depends largely on the water source. The more that groundwater adds to the total streamflow, the higher the content of dissolved minerals. Small rivers have widely varying chemical content. Large rivers have similar chemical content from river to river. In North America, streams draining into the North Atlantic, the Pacific, and the Great Basin of the West carry more of their load in solution than in any other way. These rivers carry a small part of the total load carried by North American streams.

### Suspended Load

Rivers carry a portion of their load in suspension. Friction of the water around the particles keeps them in suspension. Material carried in suspension depends primarily on the velocity of the stream and the size of the particles entrained. The finest clay particles stay in suspension if there is any motion in the water at all. The Missouri River, for example, is called the Muddy Missouri or Big Muddy. The Missouri River flows through areas where the bed is made up of very fine clay-sized material. Particles stay in suspension if the upward frictional force around the particle is greater than the pull of gravity on the particle. Since water is more dense than air, larger particles stay in suspension in water than in air. You can demonstrate this by throwing a handful of sand into a clear pond or stream. The sand falls rapidly through the air until it strikes the water. When it hits the water, the velocity begins to slow until it reaches a steady rate of fall. Whenever the turbulent velocity is greater than the fall velocity of the particles, they stay in suspension. The total amount of material carried in suspension by a stream varies with the volume of water (discharge). The greater the volume of water, the more sediment there will be. In North America rivers carry about two-thirds of their load of some 4 billion tons each year in suspension.

Other material moves as bedload. Bedload consists of particles moving along the bed of a stream by sliding, rolling, or bouncing (saltation). Loose particles on the bottom move when there is a sudden increase in velocity, or they move if struck by other particles. Once in motion, large particles move more readily than small ones. This is because they reach higher into the zone of turbulence and their momentum is greater. Round particles move more easily than do rectangular or irregularly shaped particles.

The amount of material moved by a stream depends on the velocity of the water and the size of the particles on the bed. Since particle size stays fairly constant over time, changes in erosion rates are due to changing velocities. Streams have their highest velocities during periods of flood, and move the most sediment during floods. The Mississippi River annually carries to the Gulf of Mexico an estimated 340 million tons of material in suspension, 40 million tons as bedload, and 136 million tons in solution. This is not a typical distribution of transported material.

The normal relationship is for most to be carried in solution and the least amount of material as bedload.

## DEPOSITION OF SEDIMENT

Streams deposit their suspended load and bedload whenever the velocity and volume drop below that needed to carry the particles. Velocity decreases where the stream enters another water body. Velocity may also decrease if the channel becomes wider and shallower. Diminished volume results from a decrease in inflow from precipitation or loss of water by seepage or evaporation. Both of the latter processes operate in areas of ephemeral and intermittent streams. A loss of water can also occur from excess use of water by phreatophytes. These are plants that use very large amounts of water and grow along streams in arid and semiarid areas. Of the 4 billions of tons of sediment carried each year, three-fourths drops to the bed of the stream or is deposited on the floodplain. The remaining 1 billion tons goes into the sea.

During deposition of sediment there is often a grading or sorting of the material. Since bedload consists of the heaviest material, it is deposited first. Often, the deposition will be in a well-sorted and stratified fashion with progressively finer particles downstream. If the velocity decreases enough, the suspended load is also deposited.

### Floodplains

A variety of distinctive landforms result from stream deposition. Some of these deposits occur along the stream channel and some at the stream mouth. The floodplain is the most common feature associated with streams. They form from the deposition of sediment (alluvium) along the stream channel when the stream is in flood. Floodplains vary from a few meters to as much as 100 kilometers in width (Figure 10.3). As the water rises and spreads out during a flood, the velocity drops. The most rapid drop in velocity occurs right next to the normal channel, and it is here that maximum deposition occurs.

### Natural Levees

Rapid deposition just outside the main channel leads to the formation of natural levees. Natural levees are low ridges formed adjacent and parallel to the main channel. On large rivers these natural levees may reach several meters in height. They range up to 100 meters or more in width at the base. They are most often visible during a flood, as they are the last areas to flood. Back from the stream the finer materials are deposited. These fine materials add minerals and a thin layer of fertile mud to the backswamp land. When floods start to recede, velocity drops still more, and deposition occurs both in the streambed and along the channel.

### Artificial Levees

Artificial levees range up to 8 meters or more in height and width. The primary purpose of the artificial levee is to prevent the river from spreading out over the floodplain that the river has built. If the water in a channel cannot spread out laterally, it can only increase in depth. Thus in many ways artificial levees are self-defeating. To decrease flooding on one area, levees necessarily raise the height of the water someplace else. Fortunately, most artificial levees will not stand the force of an extremely high flood; they break and allow the stream to occupy the natural floodplain. There

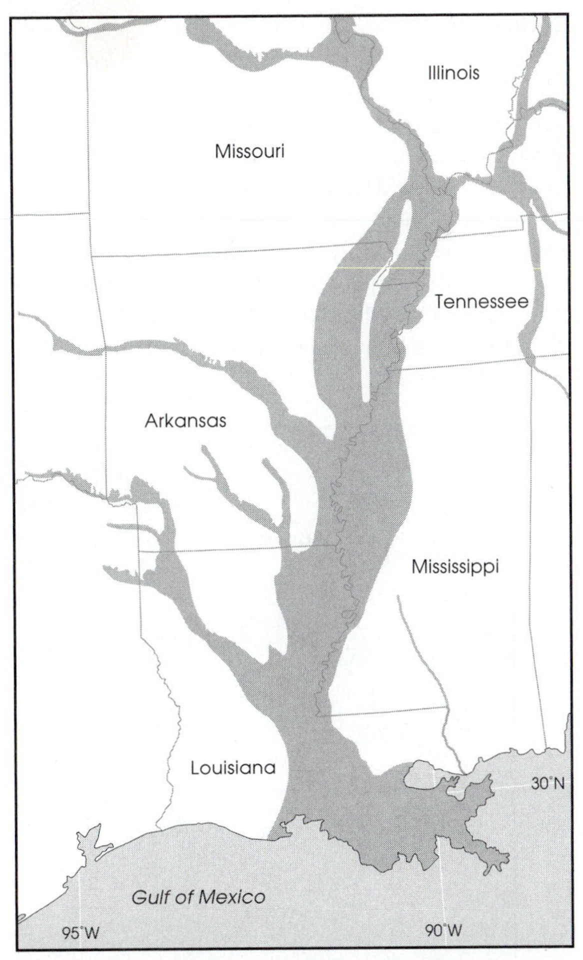


**Figure 10.3** Floodplain of the Mississippi River. The valley was cut by the meltwater of the Wisconsin ice sheet. It has since been filled with sediment eroded from the drainage basin. (From D. S. Owen and D. D. Chirsos, *Natural Resource Conservation*, Prentice Hall, Englewood Cliffs, N.J., 1995, p. 87.)

are undesirable effects of artificial levees. First, the flood cannot spread out of the channel to deposit the suspended load. Therefore, it must deposit much of the load on the channel bottom. The result is to raise the elevation of the channel bed. This increases the height to which the next flood will rise. Along one stretch of the Mississippi River, channel deposition was raising the flood level 60 millimeters per year. As flood levels rise, the area inundated increases. Second, the rise in stream elevation increases the height of the groundwater table and thus the area of backswamp.

When deposition takes place on the floodplain, it often buries crops, changes the texture and fertility of the soil, and alters the existing drainage network. The deposition of sediment on floodplains may also be beneficial. Before construction of the High Aswan Dam in Egypt, the annual enrichment of the Nile Valley was a major benefit. Floods deposited a layer of 100 to 150 millimeters of sediment on a floodplain in Nebraska. Corn yields increased by as much as 45 percent over a three-year period. But there is no doubt that the damage caused by the misplaced sediment far outweighs the benefits.

## Deltas

Where the velocity of a stream drops as it enters a lake, reservoir, or sea, deposition takes place. The rapid deposition builds landforms called deltas. Deltas can be of any size, from a few centimeters across to tens of kilometers across. The name comes from the shape of the deposit made by the Nile River: a Greek capital letter delta, **Δ**. Deltas actually may take a wide variety of shapes. The actual shape of a delta depends on the nature of the coastline. Some deltas grow very slowly and others quite rapidly, depending on the amount of sediment they carry.

The Mississippi River carries 2 million tons of sediment into the Gulf of Mexico daily. Some parts of the Mississippi River delta move seaward 1 kilometer every 10 years. Several distributaries have been closed to prevent the river from abandoning the channel past New Orleans (Figure 10.4).

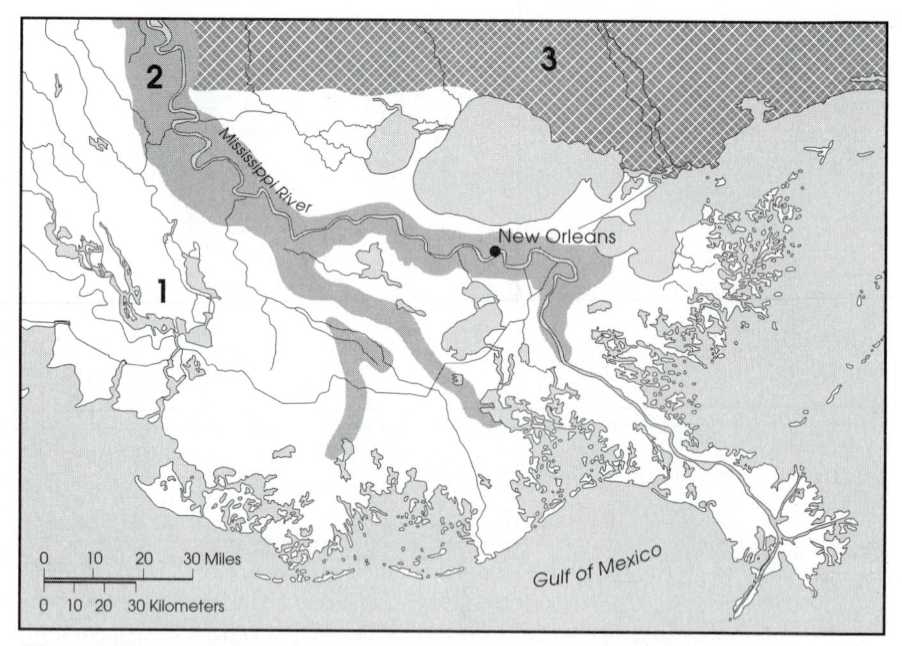


**Figure 10.4** Mississippi River Delta. The delta continually grows seaward as a result of deposition of sediment. Distributaries form and carry water during periods of flood. They often provide shorter routes to the sea than does the main channel. (1) Low-lying back swamp; (2) natural levees and higher deposits of alluvium; (3) coastal plain of older sediments.

## Impact of Sediment Deposition

The deposition of sediment causes damage in a variety of ways:

1. By filling navigable channels and harbors
2. By reducing storage space in reservoirs
3. By filling irrigation and drainage ditches

In the United States, the highest costs associated with deposition are those for dredging sediment from navigable channels and harbors. The federal government maintains a 3.6-meter navigable channel in the main channel of the Mississippi River and a 2.7-meter channel on the upper river and its tributaries. The Corps of Engineers removes some 35 million cubic meters of silt and sand from these channels annually. This is enough silt to fill 2 million average-size railway cars.

The Mississippi River deposits about 40 percent of its sediment in the river system and 60 percent in the delta region. If the material deposited in the channel system were deposited equally between the levees, it would raise the bottom of the channel by more than 25 millimeters per year. It is not deposited evenly or all within the normal channel. During the years prior to the Great Depression, the flood height for a given discharge was rising at nearly 100 millimeters per year at some locations. During the four years 1929–1933 at Hitchman Kent and Friar's Point, the bottom rose 1 meter. This elevation of the channel bottom has a number of adverse effects. It increases flood heights and total area inundated; it causes a rise in the groundwater table, which increases the extent of area inundated; and it blocks the channel for navigation.

## Sedimentation in Lakes and Reservoirs

Sediment reduces storage space in streams, lakes, and reservoirs (Figure 10.5). Large reservoirs on the Mississippi River reduce water velocity. Much sediment precipitates out where the river empties into the upper reaches of the reservoirs. The water never becomes really clear because the circulation in the reservoirs keeps the fine material in suspension. The Missouri River Division of the Corps of Engineers reports that before completion of the six major dams, 175 million tons of sediment each year went past Omaha, Nebraska. Now only 40 million tons flow past the

**Figure 10.5** Braided stream choked with sediment from mine tailings near Silverton, Colorado.

site each year. The remaining 135 million tons are deposited in the reservoirs. The large amount of sediment will make them essentially useless as storage reservoirs within a few decades.

The design of large reservoirs includes volume for sediment deposition. Sediment deposition into reservoirs in the United States is as high as 185 million cubic meters each year. Deposition in small reservoirs is higher than in large reservoirs. Nearly 40 percent of the reservoirs in the United States are under 123 cubic meters in size. Loss of storage in these reservoirs takes place at an average rate of 3 percent per year. The rate of sedimentation in small reservoirs is highly variable, depending on immediate surroundings. The watersheds above small reservoirs are small. They may be all cultivated land or completely forested. The larger the reservoir, the lower the average percentage rate of storage loss. Larger watersheds have more varied topography and vegetation cover, which moderates the rate of sediment loss. In the largest reservoirs storage loss is less than 0.5 percent per year. Although the percentage loss in large reservoirs may be low, actual volumes are large. The Colorado River deposits an average of 700,000 metric tons of silt into Lake Mead each day.

Developing countries built many large reservoirs to store irrigation water. The projected life of these reservoirs is often a century or more. A high rate of siltation reduced the useful life for many of them by 25 years or more.

## EROSION IN THE UNITED STATES

Early settlers along the eastern seaboard found the soils to be quite productive. They cleared fields of the forest and planted fields year after year. By the time of the Civil War, crop yields were beginning to decline. Soils in New England were becoming severely eroded in some areas. Farmers began to leave the New England farms and move westward.

The southern Piedmont is one of the most heavily eroded areas of the United States. An average depth of 178 millimeters of topsoil has been eroded from the Piedmont in Alabama, Georgia, South Carolina, North Carolina, and Virginia. In some areas the erosion proceeded to gullying so severe that agriculture had to be abandoned. The study showed that erosion in the area was minimal in the centuries prior to cultivation. Early reports from inhabitants uniformly agree that the streams were clear when the area was first settled, even during times of high water. The Piedmont was initially settled around 1700 and clearing proceeded to spread southwestward over the entire area. Land was cleared and farmed until it became exhausted and unproductive, and then was abandoned. As land was abundant and cheap, there was little economic reason for conservation measures. The plantation form of agriculture is certainly a major factor in the destruction of the soils. By the time of the Revolutionary War, erosion was already severe in the Piedmont portion of Virginia. The combination of corn, cotton, and tobacco left the soil bare and exposed to the heavy rains for over eight months of the year. The process spread and continued until the Civil War, during which much of the land was abandoned. Then followed the most severe period of erosion in American history, continuing until the conservation movement of the 1930s.

## NORTH AMERICA: DROUGHT AND DUST

Wind is a major erosional agent, as is water. The primary difference in the erosive power is due to the difference in density. Wind can equal or exceed water in its destructive force. This is especially true on land devoid of plant cover and where the soil surface is fine particles of silt or clay size. Wind erosion is particularly effective on dry soil. Dry soil erodes 33 percent faster than soil that has enough moisture to sustain plant growth. Wind speed and direction are important factors in determinng erosion rates. The ability of wind to move material inceases at a geometric rate with wind speed. Winds over the Mississippi River Valley can move 1000 times more soil than the Mississippi River.

In southwestern United States and northern Mexico there was a prolonged and severe drought from about A.D. 1276 to 1299. Tree-ring analysis shows that the driest year occurred in 1286. The severity of the drought was such that occupants abandoned settlements in southwestern Colorado, including Mesa Verde, and others in Utah and Arizona. Building in the Four Corners area stopped by 1285.

Droughts affected early colonies in Canada and the United States. Drought hit New England in 1749. The spring months were unusually dry and pastures dried up and caught fire. The smaller rivers dried up. During the summer conditions got still worse and drought affected most crops. There was another severe drought in New England in 1762. It was similar to that of 1749, but it was especially severe in eastern Massachusetts. The drought was severe enough that much of the livestock was killed in the fall as there was not enough feed for the winter.

The Great Plains stretch through parts of 10 western states. The soil is rich, but available water limits agriculture. Wind velocities are high, averaging 15 miles per hour year round in Oklahoma City. The Great Plains of North America suffers a chronic problem of drought. The 40-year period between 1825 and 1865 was one of low rainfall in many parts of North America. It was during this period that the Great Plains was explored but not yet extensively settled. Tracks of pioneer wagon trains exist on the floor of some western lakes. The lakes were dry in the 1840s when the tracks were made. The lakes refilled and covered the route until the turn of the century. Eighteen sixty was probably the driest year in the plains states since settlement took place. Kansas, Missouri, Iowa, Minnesota, Wisconsin, and Indiana were all affected. Drought struck the same area again in 1863–1864.

The Homestead Act of 1862 provided an incentive to move west. The act was passed to encourage settlement of the western states. Following the Civil War, thousands of families elected to move west to start new lives. A flood of pioneers into the Great Plains took place. The Homestead Act provided 160 acres of land to a homesteader at no cost if they would live on the land and farm at least part of it for five years. The Homestead Act resulted in large areas of the West being settled in the 1880s and 1890s.

The Homestead Act limited an individual holding to 64.8 hectares (160 acres) and required that the homesteader plow a portion of each holding. One hundred sixty acres would not support a family based on grazing. The result was the destruction of millions of hectares of buffalo and grama grass. Grass had served as a natural cover for the land. Plowing destroyed part of this cover and overgrazing destroyed the rest.

A severe drought peaked in the fall of 1881. It spread over all of the United States and southern Canada east of the Mississippi River system. Many of the wells, cisterns, and springs that failed had never been dry before. Lack of water for steam delayed freight trains and the water supply of New York City failed. Again, in the middle of the 1880s rainfall diminished and it culminated in the severe drought of 1894 and 1895. This was perhaps the worst drought experienced since colonization in duration and size of the area affected. Many settlers of the Great Plains left their land. Most moved westward looking for more favorable farming sites. Wetter years returned to the plains and people quickly forgot the drought problem. New settlers moved into the plains states in large numbers. Drought drove many of these away in 1910. This culminated the land runs into Oklahoma discussed in Chapter 9.

Following World War I, a third wave of farmers moved into the Great Plains on the heels of some wetter-than-normal years. By this time much of the Great Plains was used for agriculture crops, mainly corn and wheat. The remainder was used for pasture for cattle. When the dry years of the 1930s appeared on the scene, a real disaster took place. It was a disaster for the economic system of the Great Plains, the land, and the social structure. The drought of the 1930s was not of equal intensity over the entire plains nor through the decade. It was a dry period for most of the United States and southern Canada. Drought was widespread over the Northern and central plains, and the Northern Rockies. The core areas were Kansas and the Dakotas. The years from mid-1933 to early 1935 were the worst.

As the soil dried out, it began to blow. Topsoil blew away by the billions of tons. Dust storms of 1934 and 1935 became visible evidence to people as far as the Atlantic coast that severe problems existed in the west. Dust blew across the continent. Huge dust storms developed along cold fronts and moved eastward all the way to the Atlantic Ocean—thus came the name *Dust Bowl*. On May 11, 1934, a single dust storm dropped 12 million tons of topsoil on Chicago. The states east of the Mississippi River disappeared in a fog of dust. The fog resulted from the deposition of 319 million metric tons (350 million tons) of dust caught up in the jetstream and carried eastward.

The dust was so thick and so fine that it found its way into homes. People often left tracks on the floor as they walked about their homes. The dust affected transportation networks at times. Visibility was reduced to such an extent that airlines could not fly. Trains had to stop because the engineer could not see the tracks ahead. Automobiles were abandoned along highways. In the Great Plains dust blew into great drifts, sometimes burying fences. The soil loss was so great and the economic times so bad that many farmers left their farms and moved elsewhere. By 1935 an estimated 80 percent of the land in the Great Plains was suffering from erosion, and an estimated 150,000 people moved out of the plains states (Figure 10.6). John Steinbeck's novel *The Grapes of Wrath* dramatized the impact of the drought.

Two more periods of intense drought occurred in 1936 and 1939–1940. The 1936 period was very intense but short, and the 1939–1940 spell was long but not as intense. The agricultural disaster pointed out the shortcomings of the 64.8-hectare (160-acre) allotment provided by the Homestead Act.

Federal farm bills passed under the New Deal in 1933 helped the prairie farmer. Measures undertaken included the planting of shelter belts, soil and water conservation programs, grazing controls, production controls, land payments, farm credit, and federal purchase of marginal land.

Drought struck the Great Plains again in the 1950s. The flow of some perennial streams decreased to nearly half their normal flow and salinity tripled. Wind erosion was again a problem, but it was not as severe as in the 1930s. In 1957 wind damaged an estimated 13,000 square kilometers of land in the Great Plains. Another 118,000 square kilometers was susceptible to blowing. This drought ranks among the three worst on record, yet the impact on the population was far less than that of previous droughts. Different methods of agriculture aimed at conserving moisture and reducing wind erosion coupled with federal aid programs prevented the kind of chaos that was characteristic of the 1930s.

There were still problems, but nothing on the scale of the previous major drought. The net worth of farmers and ranchers declined as debt increased. Off-farm employment increased, the ratio of cattle to sheep declined, and irrigated land increased. Probably the most significant outcome of the drought of the 1950s was a migration of people out of the affected area. There was a consolidation of farms and they grew larger in size and fewer in number. There were fewer inhabitants when the rains came again.

The devastation in the Dust Bowl of the Great Plains showed the need for legislation and action to reduce the problem. With the New Deal politics of the 1930s, the Soil Erosion Service was begun. It was formed in 1933 as a temporary agency within the Department of Interior. One of its first tasks was a national survey of soil erosion. The survey found that once productive farm-

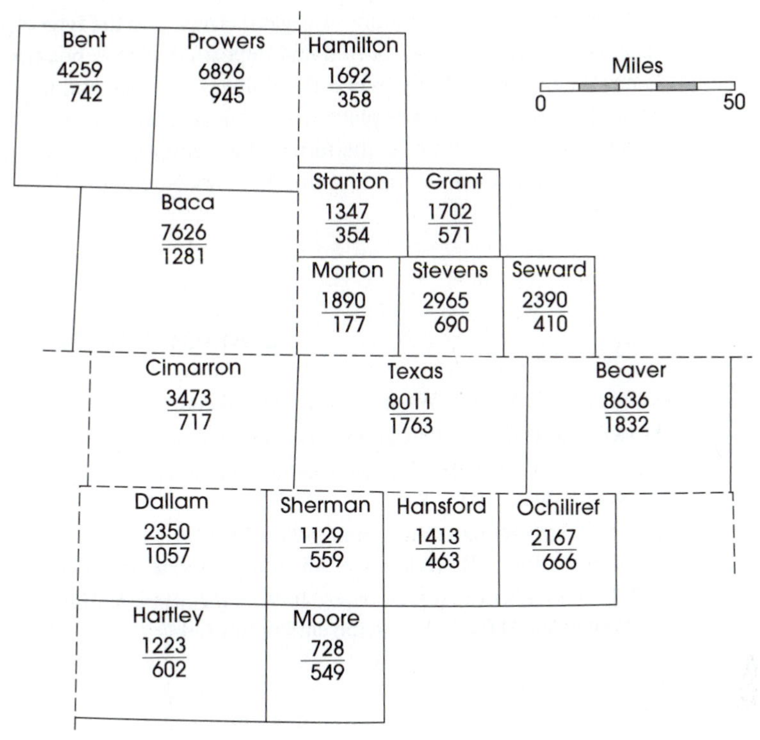


**Figure 10.6** Rural farm populations in 1930 (top number) and rural farm population in 1980 (bottom number). In most of these counties, 1930 was the census year of maximum rural farm population. Source (From M. E. Lewis, The national grasslands in the old dust bowl, Ph.D. dissertation, University of Oklahoma, Norman, Okla., 1988, p. 113.)

land had been abandoned in many parts of the country due to severe erosion. The survey estimated that erosion damage amounted to billions of dollars per year.

The Soil Conservation Act of 1935 set up the Soil Conservation Service. Millions of dollars have been spent on soil conservation in this country since the dust storms of the 1930s. However, erosion of croplands by wind and water is still a severe environmental problem facing the nation.

The primary factors of climate—temperature, precipitation, and wind—have not changed significantly over time in the Great Plains. There is a drought problem in the region, one that has widespread economic effects. The basic cause of the problem is the failure to adopt a form of agriculture compatible with the natural environmental system. The total or partial loss of agricultural products is the most widespread effect of drought. The loss is reduced yield or reduced market value of the harvested crop. In both cases there is a loss of income to the farmer and usually higher prices in the marketplace.

## SUMMARY

The process of soil erosion involves the entrainment, transportation, and deposition of sediment. Human agriculture greatly increases the rate of erosion by removing the vegetal cover from the land. The amount of soil lost to erosion depends on several factors, including the length and steepness of slope, the amount of vegetal cover, and the type of soil. Once removed from the surface, the sediment is transported and deposited. The deposition of sediment can be a boon or a bane. When spread thinly over a floodplain the added soil may enhance agricultural production. Too often, deposition takes place where it is not desired. It fills reservoirs, raises streambeds, and buries agricultural land. At the turn of the century, soil erosion is a major environmental problem, as large amounts of agricultural land become unproductive each year due to excessive soil erosion.

## BIBLIOGRAPHY

McLellan, S. M. 1993. Weathering and global denudation. *Journal of Geology*, 101:295–303.

Morgan, R. P. C., ed. 1986. *Soil Erosion and Its Control*. New York: Van Nostrand Reinhold.

Phillips, J. D. 1990. Relative factors influencing soil loss at the global scale. *American Journal of Science*, 290:547–568.

U.S. Department of Agriculture. 1991. *The Universal Soil Loss Equation with Factor Values for North Carolina*, 4th ed. Raleigh, N.C.: U.S. Soil Conservation Service

U.S. Department of Agriculture. 1994. *Soil Erosion by Water*. Agricultural Information Bulletin 513. Washington, D.C.: U.S. Government Printing Office.

chapter 11

# Global Land Degradation

***CHAPTER SUMMARY***



The Tigris and Euphrates rivers in the Middle East start in highlands of Turkey. The area is largely covered by sedimentary rock. As a result, the streams carry dissolved minerals and also a heavy load of silt. In their lower reaches they flow through the semiarid land that was Mesopotamia. Annual flooding supplies both nutrients and water to the soil of the floodplains. As the Sumerian civilization progressed (Chapter 6), more grazing and agriculture took place in the upper reaches of the Mesopotamian rivers. An increasingly larger social structure was needed to keep the irrigation systems operating and free of silt. They had to remove an increasing amount of silt from the canals. The silt then had to be spread over the fields or formed into dikes along the canals.

By 2100 B.C., problems of salt accumulation in irrigated fields also occurred, and crop yields began to decline. "As the crops changed, Girsu's field records show that grain production per hectare also declined. In 2400 B.C., the yield was very good by our contemporary standards, 2537 liters per hectare. By 2100 B.C., it was down to 1460 liters per hectare; by 1700 B.C. it had dropped to 897 liters per hectare (Taylor and Humpstone, 1973, p. 13).

Following the Sumerian civilization, greater destruction took place in the river valleys. The headwaters of the Tigris and Euphrates rivers lie in an area of hills populated from very early times. Overgrazing, deforestation, and cultivation led to extensive loss of soil. The erosion in the headwaters increased still further the loads of silt brought down the rivers. The silt carried into the delta since 1650 B.C. has built the delta of the two rivers 300 kilometers out into the Arabian Gulf.

Sediment has added 5200 square kilometers of land surface to the delta. As the silt content of the rivers increased, an ever-larger amount of labor was needed to keep the irrigation canals clear.

Between Mesopotamia and the Mediterranean Sea rapid changes in the landscape were also taking place. Forest covered much of region and timber was being shipped to Egypt by the third millennium B.C. Agriculture expanded into the region and it reached its maximum extent under the Romans in the first century B.C. Much of the agriculture depended on irrigation. Timber was still exported during the Roman occupation. High productivity was maintained until the year A.D. 636. In 636, the Bedouins from southern Arabia overran the region and much agricultural land was abandoned.

Northern Syria is said to be the "Land of Dead Cities." The ruins of 42 cities and towns abandoned in the seventh century have been located scattered among 14 that exist at present. That the area was prosperous is shown by the presence of wine and olive presses. In the Jordan Valley, which is less than 100 kilometers long, there are ruins of more than 70 ancient settlements.

After A.D. 636 there was steady degradation of the environment in the whole region. Saladin the Kurd devastated the area of what is now Israel in the year 1189. The entire Mesopotamian civilization collapsed in the thirteenth and fourteenth centuries when the Mongols and Tartars plundered the region. They destroyed irrigation systems, which were not rebuilt, due to loss of labor and excessive silting. It is certain that the population of this area has never regained the level it was before the collapse. The population of Mesopotamia may have reached 25 million at its height. The present country of Iraq has a population of only 4 million. Iraq is larger in area than Mesopotamia was at its peak. The ancient city of Babylon is now buried in sediment.

## *NOMADIC HERDING AND SOIL EROSION*

One of the major elements in the destruction of the forests was the nomads with their herds of sheep and goats. As overgrazing destroyed the grasslands, herds moved higher into the forest. They removed trees to increase the growth of grasses. Cleared land was then heavily grazed by sheep and goats, preventing the regrowth of trees. Eventually, the vegetal cover was destroyed altogether. This cleared the way for massive soil erosion. Now much of what was originally forested hills is barren of trees and soil. Where the soil remains, an ancient grove of some 500 trees still stands, known as the Cedars of Lebanon.

By 1500 the land in Israel was a desert. The Turks conquered the region in 1506. By 1800, taxes claimed two-thirds of all farm products. To avoid having to give away most of the results of their labor, the people often abandoned farmland. Taxes continued to increase until the British replaced the Turks in 1918. By this time most of the forests had been cut because there was a tax on live trees.

It has been since the beginning of the advanced agricultural societies that parts of the earth have suffered irreparable environmental damage. The landscape has been altered beyond the point of recovery. Natural feedback mechanisms can no longer restore equilibrium to a level anywhere near that existing before irrigation technology.

As agriculture spread around the Mediterranean Sea basin and across North Africa and the Middle East, land degradation followed. There are numerous examples of the damage. North Africa was once the granary of the Roman Empire. El Jem, an ancient city in what is now Tunisia, was the center of a major agricultural region. It had an amphitheater that held 60,000 people. By the 1930s there were fewer than 5000 permanent inhabitants of the region surrounding the ruins of the city.

The advanced agricultural societies represent that stage in the relationship between the human species and the environment when humans began to alter the environment to the point where it was pushed beyond the point of equilibrium. When civilization arose, food and water were stored. Negative feedback mechanisms which had restricted population growth were lessened. A major population expansion took place in association with increases in agricultural output.

The development of centralized governmental authority changed the balance of resource use. Formerly, it had been the individual family using the land to produce the necessities of life. Central authority soon began to set what each might take from the land. This inevitably led to land abuse and loss of environmental equilibrium. Resource destruction began in earnest with the more rapid consumption of resources through fishing, forestry, and mining. Indirect destruction began as well. This is the destruction of one resource to get at another. Pollution became an increasing problem, due largely to the concentration of many people in a limited area. The amount of waste was more than local environmental systems could destroy. And, of course, large-scale war, the greatest waster of resources, came into being.

Irrigation-based societies had to cope with waterborne diseases such as bilharzia. This disease is caused by a worm that spends part of its life in snails in slow-moving water. The irrigation ditches and drainage channels provided a perfect home for the snails. The recent introduction of irrigation in Aswan province of Egypt raised the incidence of bilharzia in farmers.

## GLOBAL SOIL EROSION TODAY

Soil erosion is currently the worst it has been in historic time. In nearly every part of the Earth where there is some kind of agriculture, there is accelerated erosion. El Salvador and Haiti have political, economic, and cultural structures that prevent any reduction in the high rates of erosion taking place there. Large landowners control most of the good productive land in El Salvador and grow such export crops as coffee, cotton, and sugarcane. The remaining land is too poor to provide food for the remaining population (Figure 11.1).

The situation in Haiti is similar to that in El Salvador. Wealthy farmers and North American sugar corporations own the rich valley land. This forces peasants onto the mountain slopes, which succumb quickly to erosion.

Human activity did not alter the watersheds of the Congo River or the Amazon River until recent years. Data show that for the Congo River Basin erosion is at the rate of 20 millimeters per 1000 years and for the Amazon at 47 millimeters per 1000 years. These rates approximate rates under nearly natural conditions. Both of these watersheds retained an extensive cover of natural vegetation until the last couple of decades (Figure 11.2).

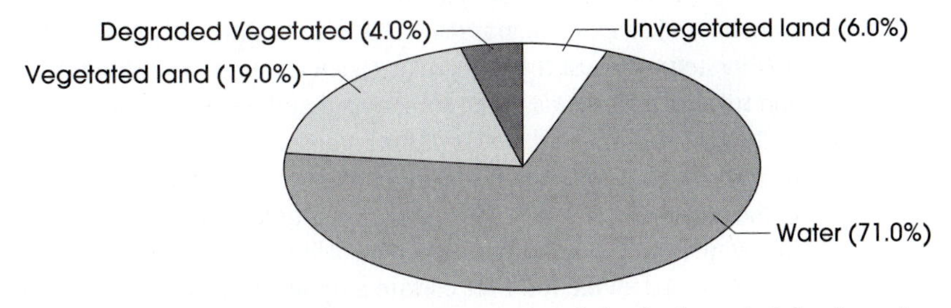


**Figure 11.1** Distribution of water, nonvegetated land, degraded land, and nondegraded land on Earth.

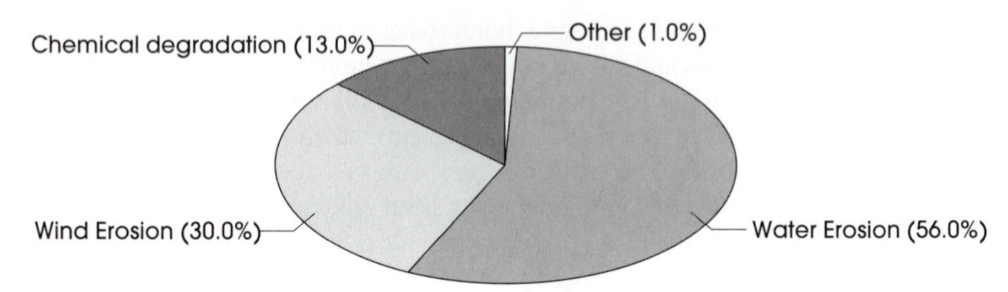


**Figure 11.2** Types of soil degradation as a percent of total degraded land (14.9 billion hectares).

From these data and other data for the United States, estimates are that an average of 36 millimeters of soil per 1000 years erode from land under natural conditions. Global rates of erosion may have doubled or tripled since agriculture began. In some areas the increase is much more than that. Total global sediment yield was about 9.3 billion tons before the beginning of agriculture and 24 billion tons in the second half of the twentieth century.

More than 3 billion acres of land, an area the size of China and Indian combined, have severely degraded since World War II. Some 22 million acres of land are so devastated that they can no longer grow crops. No country has escaped soil degradation. Two-thirds of the damaged land is in Africa and Asia. Thirty-five percent of current soil erosion is caused by overgrazing. Thirty percent of land damage results from poor cultivation practices and 28 percent results from deforestation (Figure 11.3).

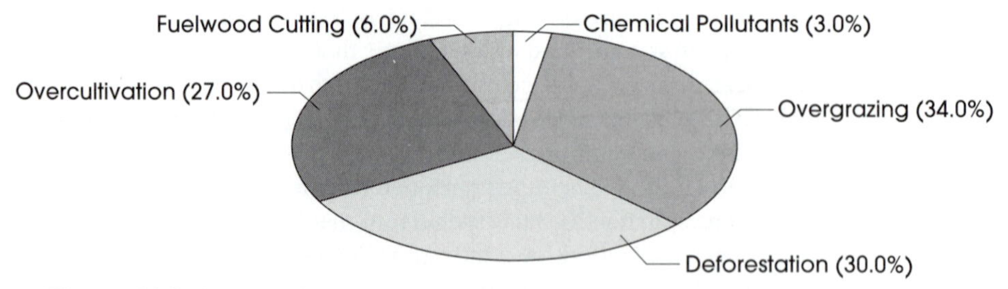


**Figure 11.3** Causes of present soil degradation in percent of total degraded land (14.9 billion hectares).

## *FIRE AND LAND DEGRADATION*

Widespread destruction of vegetation by fire has also taken place. Some fires probably resulted from natural causes, but grasslands has also traditionally been fired during the dry season. Humans deliberately set most fires. There has been a long debate about the advantages of firing woodlands and tropical grasslands. The problem is that there is a big difference between occasional and limited firing and annual burning. From whatever source, fire encourages grassland by suppressing new woody growth. When fires are hot enough or the soil dry enough, lasting damage can occur to the soil and eventually to an entire ecological system. Frequent burning encourages the growth of fire-resistant coarse grasses that have little value to grazing animals. Fires also waste the dry grasses that provide food for certain animals. Regrowth consumes considerable energy from the root zone which would otherwise be used for producing new leaf matter during the wet season.

It is the frequency and extent of firing that has done the damage. On the desert margins of the grasslands there are fewer fires and they are quite patchy, as a result of the uneven distribution of vegetal cover. Although the fires here may be less extensive in character than those closer to the equator, they are relatively more destructive. This is because the grass is all that holds the loose surface material in place.

As the length of the rainy season increases toward the equator, the volume of combustible material increases and so does the burning. Nomads moving southward during the dry season start many of the fires to stimulate regrowth of the grasses. Satellite data show scattered fires throughout the desert margins as the dry season develops. As the dry season progresses, concentrated areas of burning develop as the burn zone moves toward the equator.

Along the forest margins burning results in brush fires that consume the trees and shrubs as well as grasses. Where frequent burning takes place, only fire-resistant species survive. Fire retards growth even in these species. The forest margin itself is retreating, due partly to frequent firing. More open types of woodland replace the semideciduous forest. Along the streams, gallery forest persists because the dampness of the valley bottoms limits the influence of fire. Analysis of satellite imagery shows that in some seasons at least 20 percent of the entire area is burned. This area covers a 2600-kilometer transect from Ethiopia to western Nigeria and a north-south band 800 to 1400 kilometers wide.

## *WIND EROSION IN THE SAHARAN REGION*

Movement of dust by winds has existed around the Sahara Desert throughout historic times; it is by no means a recent phenomenon. Wind carries dust from the Sahara and adjacent regions great distances and in large quantities. This material represents the finest particles of topsoil (Figure 11.4). The dust storms called haboobs in Sudan are infamous storms. In Khartoum the frequency and severity of haboobs is increasing. Mahdi El Tom of the University of Khartoum wrote on July 25, 1975: "Yesterday we had a terrible haboob that continued for ten hours, reducing visibility to zero. The amount of dust brought to Khartoum by the haboob is unbelievable."

So great is the wind erosion from the Sahel that dust clouds float across the Atlantic Ocean on the trade winds. Measuring the amount of dust is difficult but there is evidence that it is increasing. The total drift of dust off West Africa may reach 60 million tons each year. Beginning in 1966, measurements of particulate matter in the atmosphere over the island of Barbados were begun. From 25 to 37 million tons of dust are carried westward over the Atlantic Ocean as far as Barbados each year. The type of dust material varies with time of year and wind direction. In the summer the dust is reddish brown in color, probably coming from the north side of the Sahara Desert. In the winter the dust is black, coming from the burnt-over grasslands of the Sahel. These huge dust storms moving over the ocean are visible in satellite images of the ocean. The dust travels as far as the West Indies. The fall of dust at Barbados increased from 8 micrograms per cubic meter in 1967–1968 to 15 in 1972. In 1973, at the height of the Sahel drought, the fallout of dust over Barbados reached 24 micrograms per cubic meter.

Large quantities of dust also reach the South American continent. Some 25 very large low-pressure storms form over northeastern Brazil each year. These storms draw in some 12.6 million tons of African dust. The rain from the storms spreads the dust over the soil of the Amazon Basin. The dust is high in plant nutrients and enriches the poor soil in the Amazon Basin. There is enough dust to provide about 1 pound of phosphate per acre of rain forest.

Wind erosion of surface materials is not restricted to dust particles. When large patches of bare ground form, the soil is exposed to higher wind velocities. In higher wind velocities sand-sized particles of soil move, as do soil nutrients and humus (Figure 11.5). The patches of bare ground develop into deflation pans, which are large shallow pits. Thousands of square kilometers

**Figure 11.4** Cleaning up dust following a dust storm in Khartoum, Sudan. The dust was so thick that the wind blew it into windrows in the interior of the house. Dust storms, called haboobs in Sudan, have always occurred. However, they have become more frequent and more severe in recent decades.

of land were hidden by moving sand on windy days during the drought in the Sahel in the early 1970s. Blowing sand and silt strips standing vegetation of its leaves or buries it under piles of sediment or sand dunes. The semiarid lands of Africa have an erosion rate about four times that of western Europe. Some 70 percent of semiarid land has a moderate to severe erosion problem.

## DESERTIFICATION

More than 60 years ago, Paul B. Sears wrote a classic book called *Deserts on the March*. The main point of the book is that during time of droughts, human use of the land causes deserts to expand. It appears that the change is enhanced by human use of the land. There can be no doubt that the

**Figure 11.5** Large dust devil over Ryhad, Saudi Arabia. In the heat of the day surface dust is picked up and carried long distances. Vehicle traffic loosens the soil surface, making it very vulnerable to wind erosion.

improper use of marginal land in some areas results in the expansion of desert ecosystems. The process of desert expansion is accelerating at a rapid pace. Changes that normally take decades or centuries are now taking place over a few short years. Some 6 million hectares of productive land change into nonproductive land each year. This process is now referred to as desertification. Desertification occurs when a nondesert area exhibits the characteristics of a true desert.

There are two distinct forms in which desert expansion takes place. One form is desert encroachment (Figure 11.6). The other is the breakdown of soil to a nonproductive state. Desert encroachment is a natural process due to climatic change or the migration of moving sand. This process takes place on the immediate edges of existing desert. Lands next to the desert are primarily grasslands such as the Sahel of Africa. These *savannas* or grasslands exist where there is enough rain to support drought-tolerant grasses and shrubs, but not enough to support trees.

When the distribution of rain over the deserts and surrounding areas shifts, the margin of the desert expands or contracts, depending on whether rainfall decreases or increases. Some desert expansion takes place over thousands of years. A good example is the expansion of the Sahara Desert during the Holocene. Precipitation has gradually decreased in the area since the Pleistocene, so the desert has gradually expanded. Some of the changes in precipitation take place over a few years or decades. These changes cause the desert to expand and contract continually.

The second form of desertification is a result of human use of the land. In this case the expansion is not continuous. It takes place in the form of intermittent breakdown of soils in the zone around the desert. It does not take place in the form of an even advancing front such as that

**Figure 11.6** Desert encroachment in the oasis at Al Hofuf, Saudi Arabia. The Nafud desert consists of moving streams of sand flowing south towards the empty quarter.

associated with the advance of a dust storm. It occurs most often in scattered patches of bare ground covering a few square meters up to several square kilometers. Four human activities promote this type of desertification:

1. Overcultivation of cropland
2. Overgrazing of rangeland (Figure 11.7)
3. Deforestation
4. Improperly planned irrigation schemes

The United Nations estimates that more than 27 million hectares of land are lost for agricultural production each year through desertification. The United Nations Environment Project found that more than 11 billion acres, or 35 percent of all cropland, already shows signs of desertification due to one or more of these activities.

Since the breakdown in surface cover is difficult to measure the advance of desert into grassland is not easy to measure. Some data and some estimates do exist. In West Africa during a drought in 1940–1941 some 340,000 square kilometers of land degenerated into desert. From 1964 to 1974, the Sahara Desert moved as much as 150 kilometers south into what had been grazing land. In the same area of the Sahel in 1972 and 1973, conditions were so bad north of 15° N that hardly any livestock survived and virtually no crops were harvested.

Thousands of people and millions of animals died as a result of the drought problem in the Sahel in the years 1968–1974. Untold millions of people had their lives disrupted in a traumatic

**Figure 11.7** Overgrazing and forest cover in Israel. The top photograph shows a woodland protected by a fence. Note the absence of any vegetation outside the fence. The trees are tall and well foliated. In the bottom photograph, the same type of trees growing with their roots in water but grazed by sheep on a regular basis. They are only a few centimeters tall.

fashion. The Sahel itself suffered a great deal of damage. Some of the damage natural processes can repair, but much of the damage is beyond repair.

In a recent study, the United Nations reported that the Sahara Desert expanded by some 635,000 square kilometers between 1980 and 1990 (*UNESCO Courier*, March 1993). Within this 10 years the desert expanded in some years and shrank in others. For example, between 1980 and 1984 the desert actually expanded by about 1.3 million square kilometers.

The nature of the ecosystems in much of the grasslands is such that they will not support many people based on subsistence agriculture. Dryland cultivation practices are mostly primitive and damaging to the land. Farmers clear areas and farm them for several years, during which the soil gradually deteriorates and loses fertility (Figure 11.8). When they exhaust the soil, they aban-

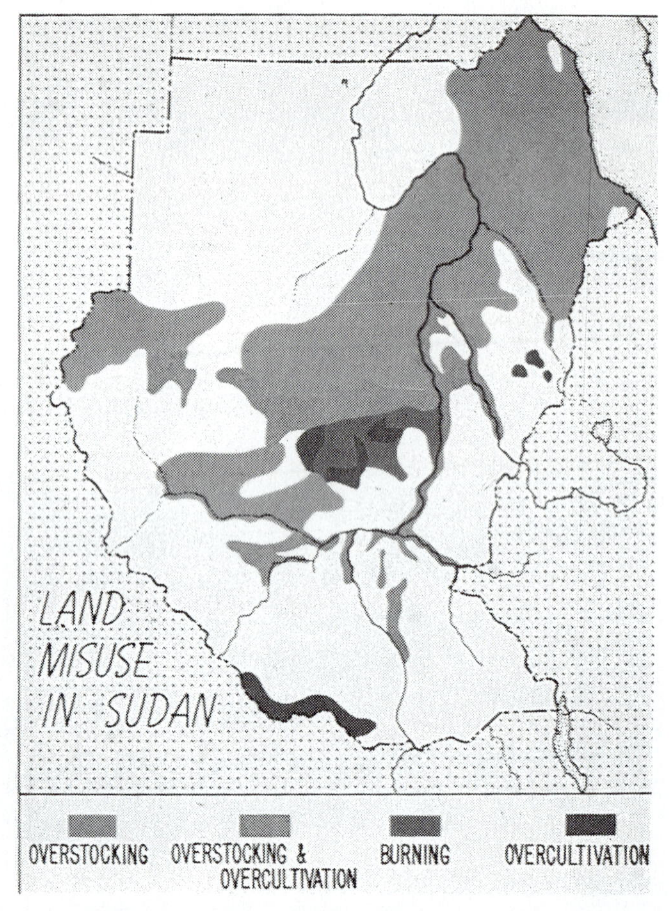


**Figure 11.8** Land misuse in Sudan. Overgrazing, overcultivation, and too frequent burning all play a part. In some regions combinations of these occur. The absence of land misuse in the northwestern part of the country is a result of the area being part of the Sahara Desert.

don it and assume that it will rejuvenate with time. The ecosystem will suffer little long-term damage if the amount of land cultivated in any given year is small enough and the livestock on the grassland does not exceed the carrying capacity.

Desertification takes place when the numbers of people and livestock exceed the capacity of the rainfall to provide food for humans and livestock. This is by no means a new problem, but it is happening at an unprecedented rate. Before colonization of Africa, intermittent periods of environmental degradation occurred. Whenever a dry cycle began, excessive grazing and cultivation led to loss of vegetation and soil. If an area became unproductive, the population either migrated away or perished. In this system the human population expanded when rain was plentiful and died back during times of drought.

The introduction of colonial administration in Africa brought some measures that altered the balance between the population and environment. Colonial governments reduced intertribal warfare and introduced public health measures. These measures resulted in rapid population growth. In 1990, several countries in the Sahel had the highest population growth rates in the world. Animal populations grew along with the human population. When the population grows rapidly in the face of varying rainfall amounts, demand on ecosystems exceeds the carrying capacity more frequently and for longer periods. Ultimately, the demand usually exceeds the carrying capacity.

The increasing pressure on the land has caused the Sahara Desert to expand southward toward the equator. Human activity may not be the only cause of the desert expansion, but it certainly accelerates it (Figure 11.9). Degeneration into desert usually occurs as patches of bare ground from a few meters in diameter up to several kilometers across. Because it is not an even front of desert surface advancing across the landscape, it is not so easy to measure. Desertification is taking place at a steady rate in many areas of Earth's grasslands and occurs on every continent except Antarctica. It accelerates during periods of major drought such as the one that struck sub-Saharan Africa in the early 1970s.

Desertification is a documented process in more than 20 countries in Africa. In the seven western countries of the Sahel, desertification is extremely great, due to deforestation. In Mali alone, the Sahara Desert has expanded 350 kilometers toward the equator since 1960.

## *SOIL SALINIZATION*

The accumulation of mineral salts is a natural process in semiarid and arid regions. It can result from natural processes or human-induced processes. Where there is not enough precipitation to percolate through the soil, water falling on the surface dissolves soluble minerals. Evaporation eventually draws the water and salts to the surface. The water evaporates, and the salts are left behind. This accumulation of salts in the surface soil is called salinization.

Humans have increased salinization by expanding irrigation systems. When not enough water is placed on the soil, so that it percolates through the root zone, additional salts come to the surface. A high concentration of natural salts such as sodium chloride and sodium sulfate has harmful effects. The salts interfere with the capacity of plants to absorb water. It also leads to the breakdown of soil structure and reduces soil aeration. This is the ability of the soil to hold atmospheric gases such as nitrogen and oxygen.

Salinization can result in two forms of degraded soils. One is saline soil and the other is sodic soil. Saline soil contains enough salts to reduce plant growth. Sodic soils have high enough concentrations of sodium to both reduce plant growth and alter the soil structure. Salinization is one means of making soils nonproductive and one of the causes of desertification. In the United States more than 3 million ha (8 million acres) of farmland have become saline.

**Figure 11.9** Surface damage from desert lorries in the Sahara Desert of Sudan. Following rains the surface softens and is deeply rutted by the trucks. These tracks then become channels for running water during rainstorms.

Salinization is an increasing problem in Iraq, India, Pakistan, China, Australia, and the Imperial Valley of California. More than a third of all irrigated land in the world has a problem with salt accumulation.

Pakistan is losing agricultural land to salinity very rapidly. The British built much of the irrigation system for Pakistan but did not construct drainage channels. Because there is a watertight layer of clay beneath the soil, excess irrigation water has no place to go. As a result, the land has become waterlogged. The groundwater table has risen to the point that about 10 percent of the irrigated land is no longer usable. The water is too saline for plants to use.

In Western Australia there was a political struggle for several decades over methods of preventing further soil salinization in the wheat belt. Salinity has increased 500 percent since the first salinization survey in 1955. Yet as late as 1970 many agricultural advisors were denying that salinity was a major agricultural problem. There is considerable disagreement about remedial measures. Many farmers, Department of Agricultural personnel, scientists, and politicians have studied the problem, but little is done.

A tragic example of soil loss due to salinization is around the Aral Sea in the former Soviet Union. Two rivers, the Syr Darya and the Amu Darya, supply water to the sea. There is no outlet, so it is a landlocked sea. This sea was once so large that ships plied the waters. There was an active fishing industry at the port called Muynak. During the 1960s and 1970s, the Soviets diverted so much water for agricultural use that the area of the sea shrank by 40 percent. The port city of Muynak is now 50 kilometers (30 miles) from water. The salinity of the sea has increased and salinization of the agricultural land has increased rapidly. The Kysyl Desert has expanded northward to encompass the sea. Uzbekistan now controls the sea and must face the problems of the region.

## *SUMMARY*

Degradation of land includes soil erosion, salinization, mineral depletion, and desertification. Land degradation began when agriculture began. The rate of degradation has increased geometrically with the growth in human population and technology. What is most disturbing is that the process continues. Severe land damage accompanies rapid development of commercial agriculture. The amount of land being lost to agricultural production is increasing at a rapid rate. Ultimately, the loss will jeopardize the ability to feed the human population. What was damaged in the past we cannot restore. At the end of the twentieth century land degradation is a worldwide phenomenon. No country is without its problems. It is unfortunate that most of the third-world countries are now going through this destructive stage.

## BIBLIOGRAPHY


BLAIKIE, P., and H. BROOKFIELD. 1978. *Land Degradation and Society*. New York: Methuen.

BROMLEY, D. 1990. The causes of land degradation along "spontaneously" expanding agricultural frontiers of the third world. *Land Economics*, 66:93–101.

DAVIDSON, D. 1992. *The Evaluation of Land Resources*. New York: Wiley.

GRAINGER, A. 1990. *The Threatening Desert*. London: Earthscan

JOHNSON, D. L., and L. A. LEWIS. 1995. *Land Degradation*: *Creation and Destruction*. Cambridge, Massachusetts: Blackwell Press

MAINGUET, M. 1991. *Desertification: Natural Background and Human Mismanagement*. Berlin: Springer-Verlag.

MEYER, W., and B.L. TURNER II. 1994. *Changes in Land Use and Land Cover*: *A Global Perspective*. New York: Cambridge University Press.

SEARS, P. B. 1935. *Deserts on the March*. Norman, Okla: Oklahoma University Press.

SIMMONS, I. 1991. *Earth, Air and Water*: *Resources and Environment in the late 20th Century*. New York: Routledge, Chapman & Hall.

TAYLOR T. B. and C. C. HUMPSTONE, 1973, *The Restoration of the Earth*. New York: Harper and Row. p. 13.

chapter 12

# Human Impact on the Hydrosphere

### *CHAPTER SUMMARY*



Most of Earth's water is in the sea and is salty. Less than 3 percent is fresh water. Most of the fresh water is frozen in the ice caps on Antarctica and Greenland and in mountain glaciers. Only 23 percent of the fresh water on the planet circulates in rivers, in the ground, and in the air. Ninety-six percent of the fresh water is in the ground. The remaining small amount is in the rivers and lakes and air. Fresh water is not evenly distributed around the Earth. Some regions, such as the deserts, have a perennial shortage of water. Much of the rest of Earth experiences seasonal shortage of precipitation and river water (Figure 12.1).

The use of fresh water increased with population growth and technological development. Total world water use more than tripled between 1950 and 1980. This increase in water use threatens both the quantity of water available for use and the quality of water.

The sea has attracted the human species for thousands of years. A large share of the world population (3 billion) now lives within 100 kilometers of the seashore. This number may double to 6 billion within the next two or three decades. Humans are moving toward the seashore for a variety of reasons. One is that people are attracted to the shore for recreational and aesthetic rea-

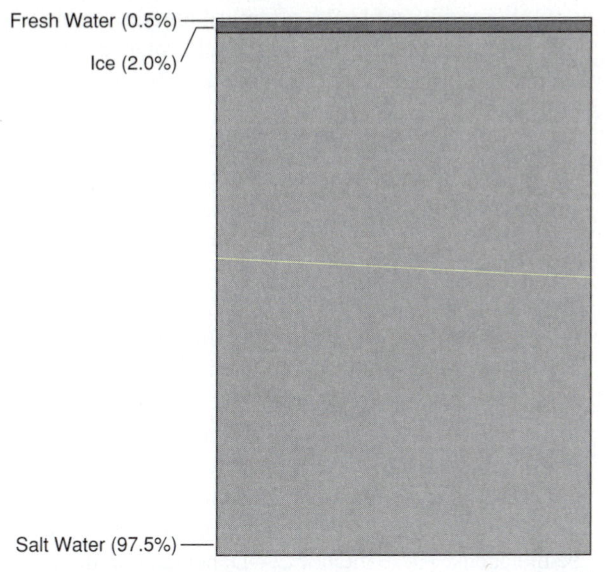


**Figure 12.1** Distribution of Earth's water supply. Almost all of Earth's water is salt water. The amount of fresh water in the atmosphere, lakes and streams, and groundwater is very limited.

sons. Another is that many of the largest and fastest-growing cities are on the seacoasts. The human species has made substantial changes in the water resources of the planet just as in the rest of the environment. These changes take many forms. There is pollution, habitat destruction, and depletion of marine life. Perhaps the most significant is change in water quality. The quality of water in most rivers, groundwater, and the sea is in various degrees of deterioration.

## *NONPOINT SOURCES OF SOIL AND WATER POLLUTANTS*

Rivers contribute three-fourths of the pollutants found in the sea. Pollutants in rivers consist of a wide variety of wastes. There is waste from agriculture, industry, urban runoff, and human sewage. Some pollutants are widely distributed and have no single source. These pollutants are nonpoint pollutants.

### Agricultural Pollutants

Agricultural chemicals are the leading pollutants in America's rivers. These pollutants include fertilizers, pesticides, herbicides, and animal waste. There has been a 600 percent increase in the use of chemical fertilizers since 1950. Nitrate- and ammonia-based fertilizer use increased by 400 percent just between 1960 and 1985. Livestock adds five times as much organic waste to water as humans do and twice as much as do manufacturing industries. Other sources of nonpoint pollution are salt from irrigation schemes, asphalt, and naturally radioactive rock. Agricultural wastes such as fertilizers and animal wastes contain phosphates and nitrates that accelerate eutrophication. Eutrophication is the process of the removal of oxygen from the water by too much plant growth. Phosphorus and nitrogen nourish the algae and cause blooms. These chemicals wash into rivers from fields and farmyards.

Some adverse health effects of nitrates are blood poisoning, hypertension in children, gastric cancers in adults, and fetal malformation. Because infants lack gastric acid, nitrate may change to nitrite. Nitrite combines with hemoglobin and forms methemoglobin, which does not transport oxygen. When this occurs, an infant suffocates. Fortunately, there have been only a few cases of infants dying of nitrate poisoning.

Iowa farmers have an increased risk of gastric cancers in counties with high corn, milk, and cattle production. This, in fact, includes most of the counties in Iowa.

### Pesticides and Herbicides

Due to their nonselectivity and the magnification process, virtually all the earth's fauna, including human beings, have assimilated pesticide residues. In the United States, most human foods contain measurable levels of residues. It has been shown that even human milk contains residues to make it a hazard for breast-fed children. The residues have spread to all parts of the Earth. They exist in the tissue of birds and seals in the antarctic, although DDT has never been used on the continent. Most synthetic organic chemicals are known to have carcinogenic, mutagenic, teratogenic, and oncogenic effects. Most known and suspected carcinogens also have the potential to be mutagens. These include 2,4-D, heptachlor, dioxin, PCBs, and trichloroethylene. (See chapter 16 for additional data on pesticides and herbicides)

## *POINT SOURCES OF POLLUTION*

### Urban Runoff

Urban runoff includes most of the kinds of materials that end up on driveways, parking lots, and streets. It includes a mix of heavy metals, oils, garbage, salt, and organic waste. Storm drainage systems often are too small to carry the volume of water present during storms. What the sewer systems cannot carry goes directly into the rivers. Many communities build storm drains to bypass sewage treatment plants altogether. Street cleaning practices help in removing trash, but chemicals seep into the cracks and crevices and wash out with the rains. In the state of Florida, street runoff is responsible for more than half of the total water pollution in the state.

### Sewage

Sewage is a large contributor to the poor quality of freshwater systems and oceans. Most developing countries lack the funds to build sewage treatment facilities. They discharge wastes directly into streams by choice. Sewage often contains harmful bacteria and viruses that can cause hepatitis, typhoid, and cholera.

### Industrial Waste

Industrial wastes became a problem with the beginning of the industrial revolution. Factories often located on rivers for one of several reasons: (1) the river provided transportation, (2) the river was a source of power, or (3) water was a major ingredient in the product or in processing the product. Factories dumped wastes directly into streams. This practice is still widespread, even in the developed countries. Sewer drains provide a hidden means of disposing of all kinds of hazardous wastes. Instances of fish with cancers and tumors resulting from metal ingestion are prevalent in heavily industrialized areas. In North America such areas include the St. Lawrence Seaway, the Great Lakes, and Puget Sound.

## SEVERE CASES OF RIVER POLLUTION

In the United States many rivers are in poor condition. The Environmental Protection Agency (EPA) issued a report in 1994 that stated 40 percent of river miles and 45 percent of the acreage of lakes it studied showed damage or was in poor condition. Wetlands are a major element in improving water quality, but the area of wetlands continues to decrease in response to urban expansion and agricultural expansion. American Rivers, an environmental group, found some rivers in very poor condition. They are the Mississippi, Missouri, Columbia, Snake, and Clarks Fork of the Yellowstone River, and Clavey River in California.

There are many examples of heavily polluted rivers in the past and the present. Three such examples are the Hudson River of North America, the Rhine River of Germany, and the Ganges River in India.

Pollution problems on the Rhine River go back as far as the early eighteenth century, when a treaty was set up to restrict transport of toxic substances. It did little good, as additional treaties in 1963 and 1976 yielded minimal results. In 1986, a warehouse owned by Sandoz Chemical Corporation and located on the bank of the Rhine in Schweizerhalle, Switzerland, burned. It took more than five hours to control the blaze. Millions of liters of water poured on the fire soon filled protective catchment basins and flowed into the Rhine. Water containing high concentrations of pesticides and other chemicals flowed into the river. The toxic mixture included 1351 metric tons of chemicals. There were 987 tons of toxic agricultural chemicals and 364 tons of nontoxic chemicals. The contaminated fluid formed a red trail down the Rhine. The red color was due to some 2 tons of mercury included in the fluid.

A week later, a leaking drainage seal under the warehouse sent an additional 30 to 60 tons of toxic materials into the river. Thousands of dead fish appeared along the banks and there was a very high percentage of fish kill. Cities in parts of Switzerland, France, Germany, and the Netherlands depend on the Rhine for drinking water. These systems had to be shut down.

Sandoz responded as positively as possible under the circumstances. They built additional holding ponds in case of future catastrophes. The local fire department monitors the chemical inventory weekly. Sandoz stopped holding products containing mercury. It is interesting that none of the countries downstream from the plant went to international court. The probable reason is that they dump freely into the river as well. A court suit would have made such information public. In 1990 the river was still badly polluted and few fish lived in the river. Governments do not lean too hard on polluters for fear the industries will move.

Among Hindus the Ganges River is renowned for its spiritual purity, but it is far from a pure stream in the physical or biological sense. The remains of partially cremated corpses, raw sewage, and industrial wastes are all visibly part of the river. More than 1 million people bathe in the river each day, and many who do contract disease. In 1986, Prime Minister Rajiv Gandhi started a 10-year cleanup campaign on the river. The program would construct sewage treatment plants and control the disposal of industrial wastes now entering the river.

The Hudson River has become steadily more polluted over the years. Industrial wastes placed in the river have increased. In the years from 1945 to 1977, General Electric plants in Hudson Falls and Fort Edward, New York released a half million pounds of PCBs into the river. The PCBs dropped to the streambed and mixed with the sediments. Bottom-feeding fish take in some of the chemicals. In 1990 fish taken from the river below the plants were unsafe to eat. Industry blocked efforts to clean the river of the PCBs. Rightfully so, General Electric believes that they will be liable for the cleanup costs. The company has advocated using methods that will cause the chemicals to break down in place.

## *GROUNDWATER QUALITY IN THE UNITED STATES*

Two-thirds of all groundwater pumped in the United States is used for irrigation. One form of irrigation widely used in the Great Plains is center-pivot irrigation. In this type of irrigation a sprinkler system attached to a pivot moves around in a circle. Most center pivot systems use groundwater. Often, irrigators add fertilizers and pesticides to the water.

Due primarily to irrigation, many aquifers are drained of water at a rate faster than it is replenished. One such aquifer is the Ogalala formation in the Great Plains (Figure 12.2). In the coastal plains of Texas the strata incline and drain toward the sea. When these coastal aquifers are drawn down, salt water encroaches into the wells. Depletion of groundwater in southern Texas allowed salt water from Galveston Bay to flow into the Ogallala aquifer. In the United States overpumping also occurs where wells supply air-conditioning systems with water.

Although water is one of the world's most valuable resources, groundwater is one of the least protected. The most common pollutant in groundwater is human and animal feces. Worldwide, there may be 10 million deaths a year due to intestinal diseases contracted from water. In the 1970s, groundwater contaminated with human feces was responsible for one third of disease outbreaks reported to the Center for Disease Control in Atlanta.

**Figure 12.2** Windmill and waterhole on rangeland in Dallam County, Texas. The windmill pumps water from the Ogallala formation.

Industries recycle, detoxify, or destroy only a small amount of their waste products. In the United States industries dispose of about two-thirds of all hazardous waste through injection wells, storage pits, underground storage tanks, or landfills. Of the total volume of waste disposed of on or below ground about 80% will appear in groundwater.

Injection wells are the most commonly used method of hazardous waste disposal in the United States. Industry injects an estimated 10 billion gallons of sewage, radioactive waste, heavy metals, chemicals, and brine into the Earth in the United States each year. Most injection wells are in Texas and Louisiana. The remainder are mainly in the midwestern states of Michigan, Ohio, Indiana, and Illinois. These wells typically pump the fluid to depths of a thousand meters or more. Once injected into the ground at this depth it is almost impossible to recover it.

Some scientists believe that liquid injected into the ground in northeast Ohio caused an increase in earthquake activity. A paper manufacturer in northeastern Pennsylvania injected hazardous waste 500 m into the ground. It reappeared in an unplugged gas well 6 km away from the injection site.

Underground storage tanks are a common means of storing commercial products such as pesticides, fertilizers, industrial wastes, and petroleum products. There are more than $2\frac{1}{2}$ million underground petroleum product storage tanks in the United States. The EPA estimates that at least 25 percent of these leak. There is no way of knowing how many abandoned underground storage tanks there are. Leakage occurs as a result of improper tank and pipe connections and corrosion. In addition to the tanks there is an additional 150,000 miles of underground pipe carrying hazardous waste in the United States. An unknown number of miles of pipeline carry sewage, natural gas, and other potential contaminants. There is considerable leakage from these as well.

Hazardous wastes were disposed of in landfills in the United States until 1980. Often, landfills are on inexpensive land such as marshes, abandoned sand and gravel pits, old strip mines, or in limestone sinkholes. About a third of all landfills are on permeable soil or rock. Although plastic liners help temporarily, about three-fourths of them leak.

Toxic industrial wastes and mining wastes are in an estimated 181,000 storage pits and ponds in the United States. Nearly 90 percent of these lie above aquifers that either provide drinking water or have the potential to provide drinking water. Less than a third have liners to prevent waste from leaking into the ground.

The EPA reported in 1988 that 74 pesticides occur in groundwater in the United States. There are some regions in the United States that are particularly sensitive to pesticide contamination: the southern coastal plain, the central Atlantic region, the Mississippi River delta, the prairie soils of the corn belt, and the Central Valley of California. One reason for high contamination in these areas is the fact that soils are quite sandy. Pesticides flow easily through the sandy soil into the groundwater. More than three-fourths of all wells tested by the Iowa Geological Survey in 1988 contained common pesticides.

Currently, there are no systematic testing procedures for pesticides in groundwater. It is difficult to set up regulations and standards for groundwater. There are a number of reasons for this:

1. Groundwater is not visible from the surface and its movement is difficult to monitor.
2. There are often no clear boundaries to a groundwater pool.
3. The length of time it takes pollutants to appear in a new location varies.

There is a growing use of septic systems in the United States in rural subdivisions and for vacation homes in coastal and mountain regions. The outflow from septic systems places biological contaminates, nitrates, phosphates, and septic tank cleaning fluids such as trichloroethylene into the unconfined surface aquifers. In 1986 the federal government estimated that there were

22 million septic systems in use in the United States. These systems discharge some 1460 billion gallons of household wastewater into the soil. Of the systems in operation studies show that perhaps 40 percent do not work correctly.

Federal legislation designed to clean up our water supply exists. The main legislative acts are:

1. Clean Water Act of 1972
2. Federal Insecticide, Fungicide, and Rodenticide Act of 1972
3. Safe Drinking Water Act of 1974
4. Resource Conservation and Recovery Act of 1976
5. Safe Drinking Water Act Amendments of 1986
6. Clean Water Act Amendments of 1987

Three agencies are primarily responsible for groundwater protection: the EPA, U.S. Geological Survey, and the Department of Interior. While each of these agencies monitors and regulates pollutants, the problem has consistently gotten worse. Even with legislation in place, groundwater protection remains largely uncoordinated in this country. States have some freedom to set their own strategies, regulations, and enforcement mechanisms. State legislation varies depending on major industries in the state. Only 17 states have enacted specific groundwater protection statutes. Forty-one states have set quality standards for groundwater, but they differ from one state to the next.

A study done in 1988 reported more than 2100 chemical contaminants in drinking water in the United States. This was 14 years after the passage of the Safe Drinking Water Act of 1974. Of these chemicals, 97 are known carcinogens, 82 cause mutations, 28 cause acute and chronic toxicity, and 23 promote tumors. Most of the rest of the chemicals have not been tested for adverse health effects.

## *CHEMICAL ALTERATION OF THE SEAS*

More than 70 percent of Earth's surface is seawater. The quality of seawater is critical to life on Earth because marine microorganisms produce from one-third to one-half of the global oxygen supply. There are distinct regions in the ocean based on the type of living organisms. At the shoreline are coastal zones such as wetlands or swamps, salt marshes, and estuaries. Off the shore is the shallow and gently sloping continental shelf. The continental shelf is a part of the landmass even though it is below sea level. At the edge of the continental shelf is the continental slope. This is a steep slope that leads down to the seafloor. Each of these zones is distinctive in water chemistry, light, temperature, turbidity, and ecosystems.

### Estuaries

It is in estuaries where the fresh water of rivers meets and mixes with salt water. Here the velocity of the river water slows and the water drops most of its suspended solids. Many things determine the potential of an estuary to environmental damage. The amount of water flowing into the estuary and the shape of the estuary determine the rate of at which the water is changed, or flushed. Small estuaries with a slow rate of freshwater flushing trap both river sediments and pollutants. As a result, these estuaries act as temporary holding tanks for pollutants. This is particularly true if there is a large human population living near the estuary or in the watershed. For the same reasons, some semienclosed seas are vulnerable to pollution: the Mediterranean Sea, Black Sea, and Baltic Sea. Estuaries with large volumes and rapid flushing dilute pollutants and carry them out into coastal waters or the open ocean.

We now know that estuaries are the very places to which we deliver the most pollutants. Most parts of the world deliver waste either to estuaries or onto the continental shelf. This is done either through delivery by rivers, delivery by the atmosphere, or direct delivery by coastal settlement.

In 1989, an estimated 16 trillion gallons of sewage and industrial waste was delivered into rivers and coastal waterways in the United States alone. This includes large quantities of mineral nutrients. NOAA found that 70 percent of the nitrogen and 60 percent of the phosphorus in estuaries in the United States is brought in by streams. The major sources of nitrogen are nitrates in fertilizers, raw sewage, and livestock feed. Phosphorus is a major ingredient in water treatment and in detergents.

The EPA studied Chesapeake Bay to determine the source of pollutants there. Most came from streams draining into Chesapeake Bay. Some came from as far upstream as Pennsylvania and New York. The study found that three rivers carried 78 percent of the nitrogen and 70 percent of the phosphorus into the bay. Most of the nitrogen came from nonpoint sources, mainly cropland. Most of the phosphorus came from sewage treatment plants. Banning detergents containing phosphates in the states of New York and Pennsylvania reduced phosphorus discharge by 29 percent between 1985 and 1988. Elevated levels of cadmium, chromium, lead, mercury, and zinc occur in coastal waters near large urban areas, including Boston, New York, and Los Angeles.

Second to the inflow of streams as a pollution source in coastal waters is atmospheric deposition. Airborne pollutants from industry, power plants, and agriculture are a major part of the total. Large airborne particles travel fairly short distances before they precipitate out of the atmosphere.

### Open Ocean

The surface waters of the open ocean are subject to considerable movement and in some latitudes have considerable amounts of nutrients. Continents are the source for all the nutrients that nourish the plankton in the ocean. It is erosion of the continents that provides the dissolved minerals needed by living organisms in the ocean.

The deep ocean is less productive than the continental shelf or coastal zone. This is because little sunlight and fewer nutrients reach the bottom. The circulation in much of the deep ocean is slow. While there are species of animals living on the seafloor, growth rates are slow and the variety of species is much less.

## *DEPLETION OF FISHERIES*

The world catch of fish has increased throughout history but is now showing signs of severe stress. Overfishing, bycatch losses, and pollution are all taking their toll on the biodiversity of the seas.

### Bycatch Losses

Bycatch losses are those of fish species taken accidentally when fishing for another species. Shrimp boats use nets to catch fish at a variety of depths. In addition to shrimp, a variety of fish are taken as well. Most of the fish are thrown back, most of them dead. The National Marine

Fisheries Service estimates that shrimp boats discard 9 kilograms of unwanted species for each kilogram of shrimp caught. The waste amounts to more than 450 thousand tons per year in the Atlantic Ocean and Gulf of Mexico. The same organization reported that in 1990 Alaskan fishermen fishing for pollack and cod dumped nearly 30,000 tons of other fish back into the sea. In 1991, 2000 tons of fish actually rotted on the docks because freezing plants could not handle them.

Drift nets are very large nets used to fish near the surface. They are suspended by floats to hang like large curtains in the ocean. Drift nets do serious damage to marine life (Figure 12.3). Many nontarget species end up in the nets. One United Nations study reports that the driftnet fishing industry in the North Pacific caught and discarded 100,000 metric tons of nontarget fish in 1988–1989. In the same period, between 300,000 and 1 million dolphins died in the nets. Another survey dealt with fishing for squid with drift nets. They sampled only about 10 percent of the driftnet operators. Observers counted 1758 whales and dolphins, more than 30,000 seabirds, 81,956 blue sharks, 253,288 tuna, and more than 3 million pomfret killed in the nets.

## Habitat Loss

Habitat destruction is an element in the reduction of global fisheries as it is in terrestrial species. Habitat reduction results from the deposit of sediment, industrial chemicals, sewage, and agricultural chemicals. Habitat loss is particularly severe in coastal systems. The coastal zone and continental shelf are biologically the most productive part of the ocean. About 90 percent of the total world commercial fish species spawn and feed in coastal zone systems. Here the water receives sunlight and nutrients. There is an abundance of phytoplankton. These are the microscopic green plants, which use sunlight to produce free oxygen and sugar. Plankton form the basis for the marine food chain.

Residues of DDT, chlordane, and PCBs exist in trace amounts throughout the ocean. These pollutants weaken the immune system of marine mammals, making them more susceptible to disease. DDT and PCBs are still a threat event, although they are no longer used in the developed countries. Since these chemicals are not subject to breakdown by bacteria, they stay in the water for a long time. Humans unknowingly eat contaminated seafood and absorb the pollutant. Some predatory fish, including swordfish and marlin, often contain large concentrations of mercury or other metals. These fish have a long life and swim with their mouths open. Because most of the pollutants collect in surface organisms, predator fish concentrate the pollutants.

An event in Minimata, Japan illustrates the effects of eating contaminated seafood. This fishing village has an industrial plant manufacturing vinyl chloride. Plant waste went into the sea. Area residents began developing illnesses in 1952. In 1956 the illnesses were traced to eating contaminated fish. There were 2000 confirmed cases and 43 fatalities from the poisoning. Many survivors had permanent health problems.

## Coral Reef Loss

Coral reefs are among the most valuable and vulnerable parts of coastal habitats. Reefs consist of small animals called polyps. Polyps and algae form a symbiotic environment in which they both thrive. Polyps deposit layers of calcareous material that eventually build into a complex reef structure. The irregular structure provides habitat for other species. Coral reefs occupy about 1 percent of the surface area of the ocean and support 25 percent of all marine species.

Reefs are threatened in several ways. There is considerable physical damage in the form of breakage. Ships run into them. Explosives used by fishermen cause breakage. There is also deliberate breaking for sale of coral for souvenirs. Pollution upsets the balance between the plants

**Figure 12.3** There is an increasing loss of marine animals and birds to trash in the ocean. In the upper photograph is a seal that perished when it got a plastic loop stuck on its snout and could not eat. In the lower photograph is a large seaturtle entrapped in an abandoned net.

and animals. In some areas silting is killing them by burying them or shutting off sunlight. Thermal pollution is also responsible for loss of coral in some areas. Polyps can live only within a narrow temperature range. An alteration of a few degrees in temperature may cause coral to die, leaving only the white bleached skeleton. Higher levels of sediment and nutrients favor the growth of algae and sponges. These organisms can eventually replace the living coral. Some 10 percent of all coral reefs have been destroyed. Another 50 percent are threatened.

### Petroleum Spills

Crude oil and its derivatives are a major marine toxin. A concentration of 0.005 part per billion kills plankton, may kill fish eggs, and may reduce the ability of fish to search for food. Most petroleum products get into the ocean from land-based sources (Figure 12.4). Runoff from streets, highways, and parking lots, and industrial waste and spills are the prime sources. The flushing of tanks in oil tankers, the pumping of bilges, and accidental spills contribute much of the rest.

A famous pollution episode was the accidental oil spill that occurred when the *Exxon Valdez* struck Bligh Reef on March 24, 1989. Despite warnings by environmental interests about the potential for an accident in Alaska's Prince Edward Sound, the oil companies did not consider it probable. They considered the chance of an accident in either Prince Edward Sound or at the terminal to be 1 in 1 million. Whatever the odds, the event happened. The oil slick that resulted from the accident eventually covered an area larger than the state of Rhode Island. It was later estimated that more than 10 million gallons of oil spilled into the sound. There was extensive loss of animal life. As many as 2000 sea otters and 33,000 birds perished from immersion in the oil.

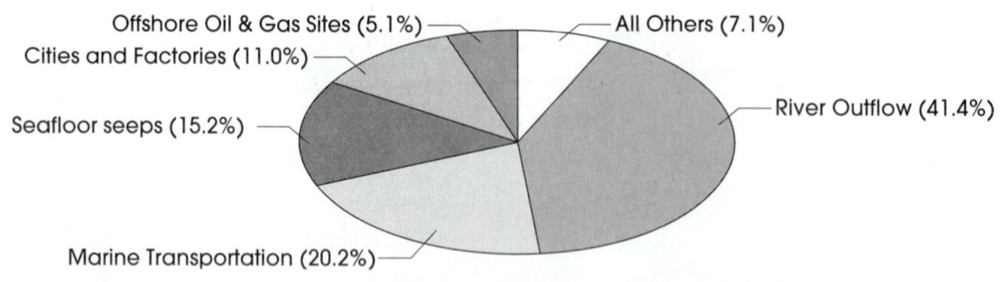


**Figure 12.4** Sources of petroleum products found in the ocean. More than half is carried into the ocean by rivers and discharge from coastal cities and industries.

## *INTERNATIONAL POLITICS AND MARINE POLLUTION*

The international record for correcting some of the major abuses of international waters is not very good. The London Convention, dating from 1972, bans the discharge of highly radioactive and hazardous waste into the sea. Out of 125 possible signatories, only 65 agreed to the convention. The Law of the Sea Treaty was written in 1982. This document seeks to establish regulations to conserve ocean resources. In 1992, 10 years after its writing, only 52 countries had ratified it. About half of the nations with international shipping fleets signed the International Convention on the Prevention of Pollution from Ships.

There are many reasons for the lack of support for these protective measures. Not all of the developed countries think that this is a major issue. Many developing countries lack funds to meet the regulations. There is almost no way to enforce the regulations.

In the United States the federal government has jurisdiction over commercial uses of the sea out to 200 miles. States have jurisdiction over land-use decisions concerning watersheds, estuaries, harbors, and shorelines. They also have certain rights over the use of water and seabed within the 5 kilometer territorial limit. States often delegate authority to local governments. The mix of federal, state, and local interests leads to inconsistent and often ineffective policy.

## *SUMMARY*

Earth's supply of water has become badly contaminated by human activity. Most of the contaminates come from the land. The ultimate sink for many of the agricultural chemicals used today is the ocean. The water draining from city streets, the outflow from sewage treatment plants, and industrial wastes also find their way into the streams, groundwater, and finally the ocean. The most contaminated parts of the oceans are the estuaries and coastal waters. Not only do rivers carry pollutants into the estuaries, but there is direct flow of pollutants from coastal cities.

Both flora and fauna in the world ocean are suffering from pollution and from fishing. Many species of fish have declined so far and so rapidly that commercial fishing has ceased. Some species have become extinct due to overfishing. Others are in serious danger due to pollutants. International politics has resulted in protection for some species but has failed to act to ensure the survival of many others.

## BIBLIOGRAPHY

BURGER, J., ed. 1994. *Before and After the Oil Spill: The Arthur Kill*. New Brunswick, N.J.: Rutgers University Press.

CULOTTA, E. 1992. Red menace in the world's oceans. *Science*, 257:1476–1477.

GORMAN, M. 1993. *Environmental Hazards: Marine Pollution*. Oxford: ABC–Clio.

HOULT, D. P., ed. 1994. *Oil on the Sea*. London: Plenum Press.

LAPOINTE, B. 1989. Caribbean coral reefs: are they becoming algae reefs? *Sea Frontiers*, 82–91.

LAWREN, B. 1992. Net loss. *National Wildlife*, 47–53.

SATCHELL, M. 1994. The rape of the oceans. *U.S. News and World Report*, June 22:64–75.

WEBER, P. 1994. It comes down to the coasts. *World Watch*, March–April:20–29.

chapter 13

# Global Changes in Atmospheric Chemistry

***CHAPTER SUMMARY***



It makes no sense in today's world to consider the atmosphere in its natural form for the reason that human activity is rapidly altering atmospheric chemistry. Technological processes inject into the air a wide variety of solids, liquids, and gases that are collectively called pollutants. It is true that the atmosphere is never completely pure under any circumstances. Gases such as sulfer dioxide, hydrogen dioxide, and carbon monoxide are continually released into the air as by-products of natural occurrences such as volcanic activity, decay of vegetation, and range and forest fires. Thus some effluents introduced into the atmosphere are in fact natural constituents.

These materials become pollutants only when they are placed into the atmosphere in abnormally large amounts. The volume of effluents placed in the atmosphere has reached the extent where levels of some constituents are increasing beyond natural limits. The EPA released a report in 1989 which indicates the magnitude of the industrial injection of chemicals into the atmosphere. In the United States there are at least 1600 industrial facilities in 46 states that release into the air significant amounts of chemicals that are suspected of being carcinogenic. Some 125 of these plants release more than 400,000 pounds of chemicals each year. There are some 30 industrial facilities that each emit more than 1 million pounds a year. A total of 2.7 billion pounds of pollutants were placed into the atmosphere in 1987. Of this amount, 360 million pounds is suspected of being carcinogenic.

## *ATMOSPHERIC POLLUTANTS*

### Particulates

In 1994, nearly 23 million Americans, 9 percent of the total population, lived in areas where particulates such as soot and acid aerosols exceed EPA standards (Table 13.1). In these areas the air often is thick and hazy, especially in the summer (Figure 13.1). Particulates can impair the functioning of the lungs and seriously threaten health. They mainly affect people with chronic respiratory illnesses such as asthma, bronchitis, and emphysema. Lung disease is the nation's third leading cause of death. California has the strictest laws governing particulate air pollution. The EPA standard for particulates is 150 micrograms per cubic meter of air. California has a maximum of 51 micrograms. The primary sources of particulates are diesel trucks and buses, factory and electric utility smokestacks, car exhaust, burning wood, mining, and construction. Small particulates have been linked to disease. Dust-sized particulates less than 10 microns in diameter have been linked to bronchitus, asthma, pneumonia and pleurisy in children.

### Carbon Monoxide

By volume the greatest emission from human activity is carbon monoxide. It is a colorless, odorless, and tasteless gas. It is formed primarily by incomplete combustion of coal, fuel oil, and gasoline. The largest single source is from the automobile. The gas begins to affect the human body at concentrations of about 100 parts per million. At this concentration people develop headaches and may become dizzy. Levels of 100 parts per million have been observed in some urban areas, and concentrations of 370 parts per million have been recorded inside vehicles trapped in traffic jams. Carbon monoxide has a residence time of several days. Eventually, carbon monoxide combines with oxygen to form carbon dioxide.

**TABLE 13.1.** Counties in the United States That Exceeded Federal Limits for Particulate Air Pollution in 1994

| | |
|---|---|
| Maricopa, Arizona | Flathead, Montana |
| Inyo, California | Lane, Oregon |
| Los Angeles, California | Philadelphia, Pennsylvania |
| San Bernadino, California | El Paso, Texas |
| Denver, Colorado | Salt Lake, Utah |
| Kootenai, Idaho | Utah, Utah |
| Cook, Illinois | Weber, Utah |
| Lake, Illinois | Spokane, Washington |

**Figure 13.1** Streams of smoke from the burning of waste gases in the oil fields of Saudi Arabia. Burning places huge quantities of carbon into the atmosphere as well as wastes valuable hydrocarbon fuels.

## Sulfur Compounds

Sulfur oxides are the second most abundant pollutant. Sulfur dioxide ($SO_2$) is one of the major oxides of sulfur. It is a heavy, pungent, colorless gas. It forms from the combination of sulfur from emissions of coal-burning industries and atmospheric oxygen (Figure 13.2). Sulfur dioxide is highly reactive and hence is not cumulative. The maximum residence time is probably 10 days. Much of the compound combines with atmospheric water to form sulfuric acid. Atmospheric sulfuric acid causes the leaves of plants to turn yellow, dissolves limestone and marble, and is highly corrosive of iron and steel. It also reduces atmospheric visibility and blocks out sunlight. It is a major irritant to the eyes and respiratory system and is lethal at a few parts per million.

In 1985 some 23 million tons of sulfur oxides was emitted into the atmosphere in the United States alone. Seventy percent of the sulfur dioxide (16 million tons) came from burning low-grade coal and petroleum in electric power plants. Sulfur dioxide emissions from electric power plants have declined since 1975 by about 30 percent due to the use of higher-grade coal and cleaner burning plants. Most of the most severe offenders are located in the Ohio River valley.

Hydrogen sulfide ($H_2S$) is another sulfur compound that forms in the atmosphere. It forms from organic decay when there is not enough oxygen present to oxidize the organic material. The main sources of hydrogen sulfide are swamps. It has a very bad smell, like rotten eggs, but fortunately has a short residence time. In the atmosphere it will darken lead in oil-based house paints. It is also responsible for tarnishing copper and silver.

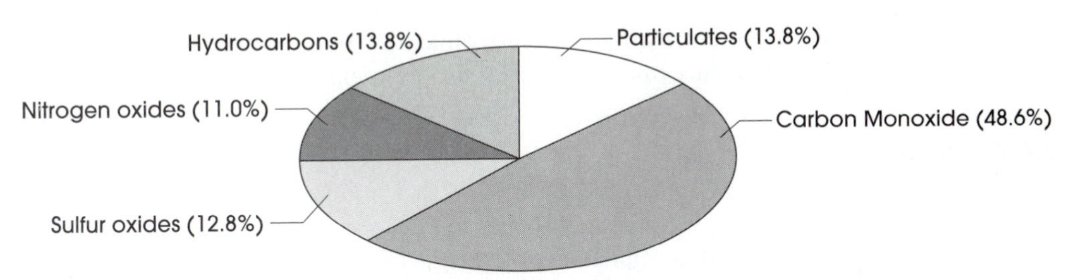


**Figure 13.2** Major pollutants emitted into the atmosphere by American industries.

### Nitrogen Oxides

Nitrogen and oxygen do not normally interact at standard environmental temperatures. The substantial quantities of nitrogen oxides ($NO_x$) result from combustion at high temperatures, largely in automobile engines. Nitrogen dioxide ($NO_2$) is the only widespread pollutant that has a color to it. It is yellow-brown in color and has a pungent sweet odor. The average residence time is about three days. The end product of nitric oxides is nitric acid ($HNO_3$). Nitrogen oxides ($NO_x$) are also a major contributor to acid rain and surface ozone. New scrubbers are being developed which remove as much as 70 percent of the gas.

Automobile and truck exhaust yields most of the nitrogen oxides in the United States. Out of a total of 20.5 million tons injected into the atmosphere in 1985, 45 percent came from vehicle exhaust. The distribution of nitrogen oxides is distributed as vehicle traffic is distributed. It is emitted over a much broader area and more evenly than is sulfur dioxide.

## *SURFACE OZONE*

Ozone is a form of oxygen that contains three atoms of oxygen instead of the usual two. The gas is colorless and odorless except at very high concentrations. It is a major ingredient in smog. Ozone is formed near the ground when pollutants such as unburned petroleum hydrocarbons and nitrogen oxides from automobile exhausts and fossil-fuel power plants react in sunlight. The chemical reactions are faster on hot sunny days. It takes several hours after the sun rises for the chemical reactions to reach the level where ozone accumulates. It usually begins to form about 10:00 A.M. solar time. The ozone that forms in sunlight usually breaks down at night. Thus the process begins each day. It does not persist for long periods.

Ground levels of ozone reached their highest levels on record in 1988. About half of the U.S. population lives where ozone exceeds the EPA standard at least part of the time. Ozone is highest in states east of the Mississippi River and in California and Texas.

Breathing ozone may cause respiratory problems for those who exercise outdoors. Ozone is an irritant to the lungs and air passages. It is especially irritating to those who engage in vigorous exercise. Some individuals are affected almost immediately if they exercise in air with elevated ozone levels. They may cough or experience chest pain and shortness of breath. To reduce the chance of respiratory irritation it is best to take precautions when exercising in warm weather when there is a risk of ozone accumulation. It is best to exercise before 10:00 A.M. Jogging along major traffic thoroughfares is also to be avoided. A general rule is the more and harder you exercise, the greater the intake of air and therefore of ozone.

The state of New Jersey took action to reduce ozone levels in 1989. They ordered the installation of new vapor control nozzles at gasoline stations. The nozzles are designed to capture the vapor given off by gasoline while it is being pumped into automobile gasoline tanks. They hope to reduce ozone levels by 3 to 5 percent.

## ACID PRECIPITATION

Acid rain is a phrase that applies to a process that results in deposition of acid on the surface of the earth. All precipitation is slightly acidic in nature. One index of measuring acidity is the concentration of hydrogen ions (pH). A neutral solution has a pH of 7.0. The lower the pH, the more acidic the water. For each unit the pH drops, the acidity increases by a multiple of 10. Thus a pH of 6 represents an acidic element 10 times that of pH of 7. A pH of 5 represents 100 times the acidity of water with a pH of 7.

Natural precipitation has a pH near 5.6. The term *acid rain* was first used by a British chemist, Angus Smith, in 1858, and refers to precipitation with a pH of less than 5.6. When precipitation has a pH of less than 5.6 it is usually due to the injection of sulphur compounds or nitrogen oxides into the atmosphere. Coal-burning electric power plants, industrial furnaces, and motor vehicles inject large amounts of these chemicals into the atmosphere. In the atmosphere the chemicals combine with water to form sulfuric acid and nitric acid. These droplets may be transported great distances by wind before they precipitate to the ground.

The acidity of precipitation has increased over North America to the level where the pH is less than 4.6 over most of the continent east of a line from Houston, Texas to the southern tip of Hudson Bay. In 1980 the pH of precipitation dropped to an average of 4.1 over part of the Ohio River valley and the Adirondack Mountains. In the Great Smoky Mountains precipitation with a pH of 3.3 was measured, and at Wheeling, West Virginia in 1980 rainfall with a pH of 1.4 was rcorded. Ordinary battery acid has a pH of 1.1 (Table 13.2).

Much of the attention on the pollution-causing acid rain in the Northeastern United States and in Canada has focused on the Ohio Valley and other areas of the Middle West where there are large concentrations of coal-fired power plants. Nine states produce 52 percent of the sulfur diox-

**TABLE 13.2** Indicator Levels of pH Related to Acid Rain

| *Item* | *pH* |
|---|---|
| Neutral Solution | 7.0 |
| Natural Rainfall | 5.6 |
| Fish Reproduction Affected | 5.0 |
| Lethal To Fish | 4.5 |
| Average Rainfall-Eastern North America | 4.4 |
| Acidity of Tomato Juice | 4.3 |
| Average Rainfall-Pennsylvania, New York, and Ontario | 4.2 |
| Rainfall at Toronto, Canada in February, 1979 | 3.5 |
| Rainfall in the Smoky Mountains of North Carolina | 3.3 |
| Acidity of Lemon Juice | 2.2 |
| Rainfall at Mt. Mitchell, N.C., July 1986 | 2.2 |
| Rainfall at Wheeling, W.Va. (1980) | 1.4 |
| Battery Acid | 1.1 |

Note for each unit decrease in the scale, acidity increases by a factor of 10.

ide emissions: Georgia, Illinois, Indiana, Kentucky, Missouri, Ohio, Pennsylvania, Tennessee, and West Virginia.

## GLOBAL DISTRIBUTION OF ACID PRECIPITATION

Acid rain has become a worldwide problem. In Europe, Norway, Sweden, Denmark, the Netherlands, and West Germany all have a problem with acid rain. In Sweden alone an estimated 18,000 lakes are more acid than natural rainfall. Great Britain is accused by the continental nations of being the main source of the pollutants. Great Britain has admitted to being a source for sulfur dioxide and nitrogen oxides.

Great Britain has also had its share of acid precipitation problems. The worst case of acid mist occurred in 1989. The mist came in over the east coast on September 9, affecting a 1000-square-mile area. The area was mainly in Norfolk and Lincolnshire. The mist was estimated to have a pH of 2.0. The mist was so acidic that it corroded aluminum instruments. It damaged thousands of trees. The acid killed the leaves, turning them brown overnight. The source of the sulfuric and nitric acid particles is believed to be automobile traffic on the continent. This incident was worse than the incident was at Pitlochry, Scotland in 1974. The Pitlochry acid mist had a pH level of 2.4, stronger than vinegar.

Precipitation in other parts of the world is acidic also. In the city of Guiyang, concentrations of the sulfate ion are about six times greater than in New York City and 20 to 100 times greater than that over Katherine, Australia. Katherine is considered to be representative of an area little affected by industrial pollution. The higher concentration of sulfates in China is due to the large use of coal as a primary fuel for home cooking, heating, and for the generation of electricity. There are virtually no controls on the use of coal as a fuel. While concentrations of sulfates are higher in China, there is a lower concentration of nitrates, due primarily to fewer automobiles. The nitrate concentration is highest in Beijing, where there are the most automobiles.

## IMPACT OF ACID PRECIPITATION

### Freshwater Systems

The impact of acid rain on aquatic environments, particularly freshwater lakes, has been clearly established. Aquatic systems are very susceptible to acidification. Fish are very susceptible to acidification. Fish become endangered when the pH drops to about 5.5. Most species of fish stop reproducing at pH levels between 5.3 and 5.6. Fish are hurt by acidification in a number of ways. As acidity increases, more trace metals are dissolved in the water. Aluminum is one such metal. Young fish are particularly susceptible to increased aluminum concentrations. The aluminum collects in their gills. In trying to get rid of it, the young fish strangle in their own mucus. Above normal acidity also prevents fish from absorbing calcium and sodium. The lack of calcium weakens their bone structure and their skeletons become deformed and are easily damaged. Lack of sodium causes convulsions that kill the fish.

Fish are also susceptible to acid shock. Acid shock is the sudden introduction of large amounts of acid. It is commonly associated with spring snowmelt in mountain regions. The acid is deposited in the snow crystals and remains on the ground for periods of up to several months. With spring melting large amounts of acid enter the streams and lakes, resulting in a sudden, if temporary increase in acidity. In northern United States, some winters, such as that of 1993–1994, have a lot of snow accumulation, and there may be widespread acid shock in the spring. Other years it may be minimal. In Canada snow accumulates in most winters, so there are annual episodes of acid shock.

By the time the pH of a lake drops to 5.0, between 30 and 50 percent of the natural biota cease to exist. The most susceptible are the smaller organisms, such as mollusks and minnows. Many lakes contain water with pH of less than 4.5. At 4.5 all fish are gone and the water supports completely different organisms from normal lake water. High acidity favors the growth of sphagnum mosses and filamentous algae.

The sensitivity of lakes to acidification depends a lot on their natural ability to neutralize the acidic runoff into the lake. Lakes located in areas where the parent rock is igneous and metamorphic, containing lots of silicates, are most sensitive to acid deposition. The dissolved minerals from these rocks result in acidic runoff. Where lakes are found in regions where the parent rock is high in the mineral salts, such as calcium, magnesium, and phosphorus, the lakes can better tolerate the acid runoff. The reason is that the soil solution tends more toward alkaline and the salts neutralize the acid.

On a global scale there may be more than 1000 lakes that have become too acidic to support life. There may be several thousand that receive episodes of acid shock. The greatest share of these are in eastern Canada and Scandinavia. Acidification is a particularly severe problem in the Adirondack Mountains of New York and New England. The parent rock in these areas is high in silicates. In these areas soils are thin and acidic under natural conditions, so the runoff from the acid rain remains highly acidic. Acidity in lakes in these areas has increased sharply since 1950.

### Terrestrial Systems

The impact on terrestrial ecosystems is less clear than for aquatic systems. A major area of controversy is whether acid clouds and acid precipitation are damaging world forests. In September 1990 the National Acid Precipitation Assessment Program (NAPAP) released the results of a 10-year study. One of the conclusions of the study was that there is now widespread forest damage in North America that can be linked directly to acid rain. Many scientists are convinced that acid rain is the leading cause, or at least the catalyst in widespread forest damage in midlatitudes. Evidence of damage to vegetation in North America is beginning to accumulate.

One area where rapid dieback of the forests is occurring is around Mount Mitchell in North Carolina. Dieback is the gradual dying of a tree, or trees, either from the crown downward or from the tips of the branches inward toward the trunk. Acid rain and acid fog are factors suggested for the problem. Fog over the mountains has frequently been measured with a pH of between 2.5 and 3.5, and rain with a pH of 2.2 was measured in 1986. The trees affected are a variety of spruce that is the remnant of a once widespread forest that was logged off more than a century ago. Dieback of this forest has been observed only since 1983. It began at the summit and has progressed down the mountain to lower elevations. Sections of the dead timber can now be seen from the Blue Ridge Parkway, which skirts the mountain (Figure 13.3).

A second area where tree damage has been documented is in New England. Stands of spruce on Whiteface Mountain in New York are dying and Hubert Vogelman, a botanist at the University of Vermont, found a 19 percent decline in the number of sugar maple trees over a 20-year period. In 1965 researchers counted 345,493 maple seedlings in a 2-acre area on Camels Hump Mountain near Duxbury, Vermont. By 1983 the number had dropped to 53,400. Vogleman also reported that wood samples show increasingly high concentrations of residues of industrial chemicals and hydrocarbons.

Forests in other parts of the world have been affected. In 1983 it was estimated that one-third of the forests in West Germany were damaged from acid rain. The extent of forests suffer-

**Figure 13.3** Balsam fir snags on Mount Mitchell, North Carolina. Trees were weakened by insect attack, acid rain, and desiccating winds.

ing from acid rain in Germany is increasing at a geometric rate. Five percent of the forests were damaged enough to be essentially dead.

It may be that the damage is done through acidification of the soil. When soils become more acid there are more dissolved metals in the soil water taken in by the tree roots and there is less decomposition of organic matter in the soil.

## Acid Rain and Health

David V. Bates, a University of British Columbia physician, found that several years of hospital records indicate that admittances increase as atmospheric sulfate levels rose in one urban area of southern Ontario containing some 6 million people. The correlation between admissions for ailments, including pneumonia and asthma, were related significantly to sulfate levels. His study indicated that 13 percent of the variations in admissions could be explained by changes in sulfate levels. Other researchers suggest that it contributes to emphysema and other respiratory diseases, particularly in children.

There may be a health hazard in eating fish taken from streams and lakes with increased acidity. Fish from these waters often have high levels of aluminum, copper, lead, mercury, and zinc. Intake of these metals can affect health. Aluminum may be linked to the onset of Alzheimer's disease. People that die of Alzheimer's disease often have high enough concentrations of aluminum to reduce neural function.

### Acid Rain and Commerce

Acid rain has affected the entry points for foreign-built automobiles in the United States. BMW of North America halted the shipment of automobiles through the port of Jacksonville, Florida in 1987 on evidence that acid rain had damaged the finish of 2000 new vehicles while they were awaiting transshipment. The rain contained a mixture of iron and sulfuric acid. The source of the pollutants is believed to be nearby power plants and soot from the stacks of ships using the harbor. At the peak period more than a half million automobiles from 20 manufacturers were offloading in Jacksonville. In response to the problem, the Jacksonville Electric Authority has been asked to modify procedures at three oil-fired plants. Shipping coming into the harbor must now use tugboats for docking and are prohibited from "blowing" soot from their stacks while in harbor. The port authority has also provided mobile washing facilities for use by automobile shippers. Acid precipitation is one of the elements in reduced atmospheric visibility in eastern United States and near large cities in western United States.

### Impact of Acid Rain on Structures

Limestone and marble are soluble in acids. Since many of the major buildings and sculpture are made of limestone and marble, they can be damaged by acid rain. In Great Britain there is a problem with acid rain in terms of the dissolving of the exteriors of major historical buildings. Forty-four of the flying buttresses on Westminister Abbey need replacement. These buttresses were rebuilt less than 100 years ago, but the limestone is badly eroded, due to solution by acid rain.

Another famous structure suffering from solution by acidic precipitation is the Taj Mahal in India. The world-renowned structure, which is built of marble, is rapidly being destroyed. Replacement of damaged panels cannot keep up with the rate of damage. On the Acropolis in Athens stands the Erectheum, a small structure away from the Parthenon (Figure 13.4). The porch roof is supported by six marble statues of maidens carved by the Greek sculptor Phidias in the fifth century B.C. The statues had to be removed because of rapid deteriortion due to acid rain. They were replaced in 1977 by fiberglass copies.

## *POLITICS OF ACID PRECIPITATION*

Pollutants responsible for acid rain may be carried long distances by upper-level winds before they precipitate out as acid rain. Acid rain traveling from the United States into Canada, has become both an environmental problem for Canada and a political problem between the United States and Canada. Canada has made a commitment to reduce pollutant emissions responsible for acid rain by 50 percent by 1995. The province of Ontario, a major source region of emissions in Canada, has adapted an emissions control plan as well.

**Figure 13.4** Statues of the maidens on the porch of the Erectheum. The Erectheum is one of the structures on the Acropolis in Athens, Greece. These statues are of fiberglass. The originals were removed to a museum because they were being destroyed by acid precipitation.

In the United States efforts to impose controls on coal-fired power plants have been opposed by the electric power industry, the coal industry, and the legislatures of the states most responsible for the emissions. They are opposed to controls, since it would cost large sums of money to install remedial technology, and the cost would need to be passed on to the consumer, raising prices for electricity. In 1986 the president of the United States endorsed a $5 billion five-year program to study cleaner methods of burning coal. Half of the cost was to be borne by the federal government and half by private industry. Nothing was done about the proposal, but due largely to prodding by Canada, the Reagan administration prepared another proposal in 1987 which would spend $2.5 billion to demonstrate technologies for controlling sulfur dioxide and nitrogen oxide emissions from coal-fired plants. More recent attempts to pass legislation reducing the levels of sulfur dioxide emissions have failed.

The U.S. Congress passed the Clean Air Act in 1970. It is of note that this was also the year in which the first Earth Day was held. The act has been revised on several occasions. The 1990 Clean Air Act Amendments specifically designated acid rain as an area to be addressed. These amendments set up a program to reduce pollutants that cause acid rain. The amendments call for cuts in sulfur dioxide emissions to a level of about half that of 1980. This means a reduction of about 10 million tons by the year 2000. Nitrogen oxide emissions must be cut by 25 percent or 2 million tons by the year 2000.

The reduction in emissions is to take place in two stages. The first targets midwestern power plants, which are the largest source of pollutants. Beginning in 1995, 111 of the most polluting coal-fired plants spread over 21 states must begin to reduce sulfur dioxide emissions. The plants that are targeted first are those that produce more than 100 megawatts of electricity per year and have a high pollution-to-power output rate. Permits will be issued that will allow a fixed amount of pollutants. This will then be cut in half beginning in the year 2000. In the second stage, beginning in the year 2000, another 200 plants must comply with reduced pollution levels. This phase will cover all plants producing more than 25 megawatts per year of power.

The 1990 amendments bring a new concept to the control of pollutants. In the past each individual polluter was given notice to cut pollutants to a certain level. The total amount of pollutants allowed into the atmosphere was not controlled. Under the new regulations there is a limit on the total. A market for emission credits will be set up so that if a polluter reduces emissions below the level set, the polluter can sell pollution credits. This actually began to take place in 1994. In this system, individual plants can pollute at different rates. Some can continue to pollute at high rates if they can buy credits from other plants. New power plants cannot come on line without having purchased the right to pollute from other plants. The total amount of emissions is fixed, but there is some flexibility within the system, which should bring about the desired reduction in pollution at the lowest cost. Not every polluter need purchase expensive scrubbing equipment. One advantage of the new system is that it allows alternatives. Plants can switch to cleaner burning fuels, install scrubbers to remove the pollutants, or buy credits from other companies. Failure to comply will result in fines of $2000 per ton of pollutant emitted over the amount allowed.

## SUMMARY

Earth's atmosphere is no longer a natural one: it is now greatly altered by human activity. The major gases of the atmosphere are still nitrogen and oxygen, but the variety and amounts of the variable gases have changed. Large amounts of dust, soot, metals, and organic matter are injected into the atmosphere each day. Large volumes of gases such as carbon monoxide, carbon dioxide, and nitrogen dioxide are also added each day. Many of the gases or their by-products are harmful to plants and animals.

Sulfur compounds and nitrogen dioxide combine with water droplets in clouds to become acid precipitation. Acid precipitation is harmful to most ecosystems. It is harmful to human health, commerce, and structures. Control of acid precipitation is difficult because the atmosphere is so mobile. The precipitation may fall hundreds or thousands of kilometers from the place where the pollutants are injected into the atmosphere.

The problem of acid precipitation can be solved only through international cooperation and a major change in public attitudes. Although it is not a long-term threat to the planet, it is a serious global problem in need of attention.

## BIBLIOGRAPHY

Budyko, M. I., and Y.A. Izrael. 1991. *Anthropogenic Climatic Change*. Tuscon, Ariz.: University of Arizona Press.

Charlson, R. J., S. E. Schwartz, J. M. Oltmans, R. D. Cess, J. A. Coakley, Jr., J. E. Hansen, and D. J. Hofmann. 1992. Climate forcing by anthropogenic aerosols. *Science*, 255:423–430.

Howells, G. 1990. *Acid Rain and Acid Water*. London: Ellis Horwood.

Moore, T. H. 1991. *Acid Rain: New Approach to Old Problem*. Editorial Research Reports, no. 9. Washington, D.C.: Congressional Quarterly, Inc.

Park, C. 1989. *Acid Rain*. New York: Routledge.

United Nations Environment Program/World Health Organization. 1992. *Urban Air Pollution In Megacities of the World*: *Earthwatch*: *Global Environmental Monitoring System*. New York: UNEP/WHO

chapter 14

# Stratospheric Ozone and Ultraviolet Radiation

### CHAPTER SUMMARY



Ozone is a form of oxygen in which three atoms of oxygen combine to form a single molecule of ozone. Ozone normally is not abundant in the lower atmosphere under natural conditions. It does, however, form in smog by the action of sunlight on oxides of nitrogen and organic compounds. This ozone does not stay in the air for very long. It reacts with other gases in the atmosphere and changes to normal oxygen molecules.

## STRATOSPHERIC OZONE

Ozone exists in the stratosphere though the total amount is small. It is concentrated in a layer, or layers, between altitudes of 12 and 50 kilometers (7 and 30 miles). The ozone is continually formed and then removed. The process by which ozone forms is the absorption of ultraviolet radiation in the range from 0.1 to 0.3 micron in length (Figure 14.1). This absorption of radiant en-

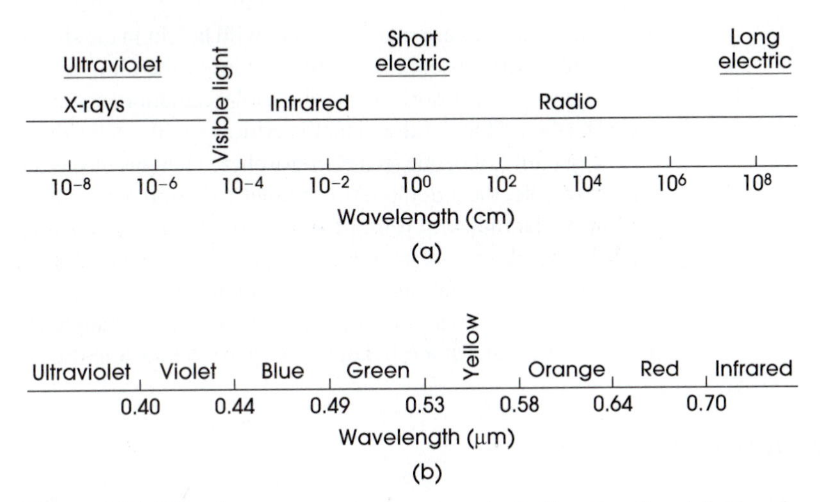


**Figure 14.1** The electromagnetic spectrum, including the ultraviolet bands.

ergy breaks oxygen molecules apart into single oxygen atoms. Some of the single atoms combine with an oxygen molecule to form ozone. Absorption of additional radiation breaks up the ozone molecules. Most ozone forms over the tropical latitudes. It is here that most solar radiation, and the most intense solar radiation, reaches Earth. The upper atmospheric circulation carries the ozone toward the poles.

The atmosphere absorbs ultraviolet radiation at all altitudes. Single atoms of oxygen absorb the shortest wavelengths (less than 0.1 micron) at altitudes above 160 kilometers. From 110 to 160 kilometers oxygen molecules absorb radiation in the range 0.1 to 0.2 micron in length. Below 110 kilometers, ozone absorbs the longer-wavelengths of ultraviolet radiation. The most ultraviolet radiation is absorbed at heights of 20 to 50 kilometers (Figure 14.2). Most of the ozone is at these altitudes. Ozone has a broad absorption band peaking at 0.255 micron. The energy ab-

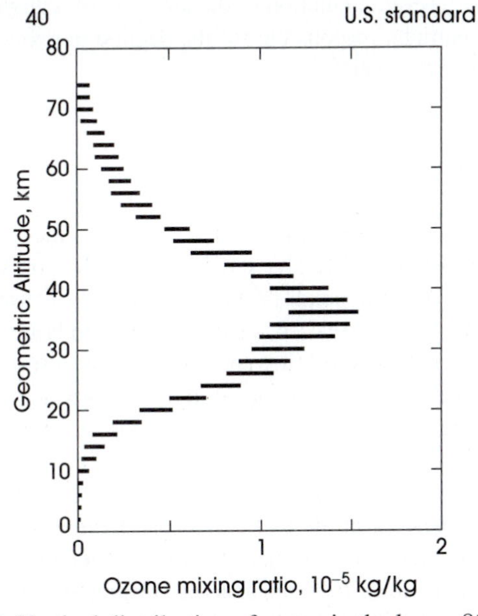


**Figure 14.2** Vertical distribution of ozone in the lower 80 kilometers.

sorbed adds to the increase in temperature with height in the stratosphere. The atmosphere, mainly as a result of ozone, absorbs about 98 percent of the ultraviolet radiation reaching Earth. While the upper atmosphere absorbs most ultraviolet radiation, some reaches the surface. The 2 percent that reaches the surface of the planet is critical to life on Earth.

The quantity of ozone in the stratosphere changes through time because of a variety of factors. For example, the amount of ultraviolet radiation reaching the planet varies with sunspot activity. Ultraviolet radiation is highest at the time of sunspot maxima and least at a time of sunspot minima. There was a peak of sunspot activity in 1979 and 1980, and it dropped to a low in 1986. As a result of these solar changes, the amount of ultraviolet radiation and ozone changes. The ozone content changes as stratospheric temperatures change. If the stratosphere warms, ozone content increases, and if temperatures drop, ozone content drops.

## *DECLINE IN OZONE*

The measurement of global ozone began in August 1931. G. M. B. Dobson set up an instrument at Arosa, Switzerland, to measure the atmospheric absorption of ultraviolet radiation. This is the longest and most complete set of data, although data are missing for a few months. Later, a network of these instruments was set up at different sites around the world. The data collected from this network represent the longest record of ultraviolet radiation and is the most reliable data available.

In 1978 the *Nimbus 7* weather satellite was launched. Among the instruments on board was the solar backscatter ultraviolet instrument (SBUV). This instrument measures the amount of solar radiation scatterred upward from the atmosphere and compares it with direct solar radiation. The satellite also carried an instrument called a total ozone mapping spectrometer.

The data show that the ozone content of the upper atmosphere has been declining in recent years. The first big change detected was after the April 1982 eruption of El Chichon, which discharged large quantities of debris into the stratosphere. It also coincided with a major disturbance in the general circulation of the atmosphere known as El Niño. Although these events may have been partially responsible for the decline in ozone, the ozone levels did not rebound once these events were over.

## *CHLOROFLUOROCARBONS*

In 1974, scientists warned that there was evidence to suggest that compounds known as chlorofluorocarbons (CFCs) have a depleting effect on stratospheric ozone layers. First synthesized in 1928, these compounds promised to have many uses. They are odorless, nonflammable, nontoxic, and chemically inert. The primary compounds are listed in Table 14.1 together with some of their characteristics. They first came into use in refrigerators in the 1930s. Since World War II, they have been used as propellants in deodorants and hair sprays, in producing plastic foams, and in cleaning electronic parts. The United States, Japan, and Europe produce and consume most of the chemicals.

Automobile air conditioners use CFC-12. This is among the most damaging of the CFCs. Ninety percent of all new cars in the United States have air conditioners. Automobiles are the largest single source of harmful CFCs. They make up 26.6 percent of the CFCs released into the environment. In 1989 the state of Vermont enacted legislation to outlaw cars with air conditioners using CFCs beginning with the 1993 model year.

TABLE 14.1 Common Chlorofluorocarbons and Halons

| *Compound (chemical formula)* | *Ozone depletion potential*[a] | *Atmospheric lifetime (years)* | *Major uses* |
|---|---|---|---|
| CFC-11 ($CFCl_3$ | 1.0 | 64 | Rigid and flexible foams, refrigeration |
| CFC-12 ($CF_2Cl_2$) | 1.0 | 108 | Air conditioning, refrigeration, rigid foam |
| CFC-113 ($C_2F_3Cl_3$) | 0.8 | 88 | Solvent |
| Halon-1211 ($CF_3BrCl$) | 3.0 | 25 | Portable fire extinguishers |
| Halon-1301 ($CF_3Br$) | 10.0 | 110 | Total flooding fire extinguisher systems |
| HCFC-22† ($HCClF_2$) | 0.05 | 22 | Air conditioning |

*Source:* U.S. Environmental Protection Agency.
[a]Ozone-depleting potentials represent the destructiveness of each compound. They are measured relative to CFC-11, which is given a value of 1.0.

These compounds are not natural compounds. They do not react with most products dispersed in spray cans. They are transparent to sunlight in the visible range. They are insoluble in water and are inert to chemical reaction in the lower atmosphere. It is for these reasons that they are valuable compounds. It is for these reasons that chemicals are a problem in the stratosphere. The average liftime of a CFC-11 molecule is between 40 and 80 years. A CFC-12 molecule may last 80 to 150 years.

Chlorofluorocarbons rise into the upper atmosphere, where they break apart under ultraviolet radiation. The breakdown takes place when the compounds are exposed to radiation of wavelengths of less than 230 nanometers. Ultraviolet radiation of this wavelength, or shorter, does not reach the troposphere because it is absorbed at altitudes of 20 to 40 kilometers (12 to 24 miles). This breakdown releases chlorine that interacts with oxygen atoms to reduce the ozone concentration. The process ends with the chlorine atom once again free in the atmosphere. Each atom of chlorine may persist for years, acting as a catalyst that may remove 100,000 molecules of ozone. The maximum rate of ozone destruction takes place at an altitude of 40 kilometers (25 miles). The final means of disposing of the chlorine is a slow drift downward into the troposphere, where it combines with water molecules and falls out as hydrochloric acid.

## ANTARCTIC OZONE HOLE

The most disturbing change in atmospheric ozone is that found over the antarctic continent, called the ozone hole (Figure 14.3). The ozone hole is a loss of stratospheric ozone over Antarctica which has occurred in September and October since the late 1970s. The hole appears in September when sunlight first reaches the region and ends in October when the general circulation brings in fresh air over Antarctica. During the antarctic spring, there is a decrease in ozone north from the pole to nearly 45° south latitude.

In August and September 1987 the amount of ozone over the antarctic reached the lowest level recorded to this date. *Nimbus 7* on September 17, 1987 recorded a large area in which the

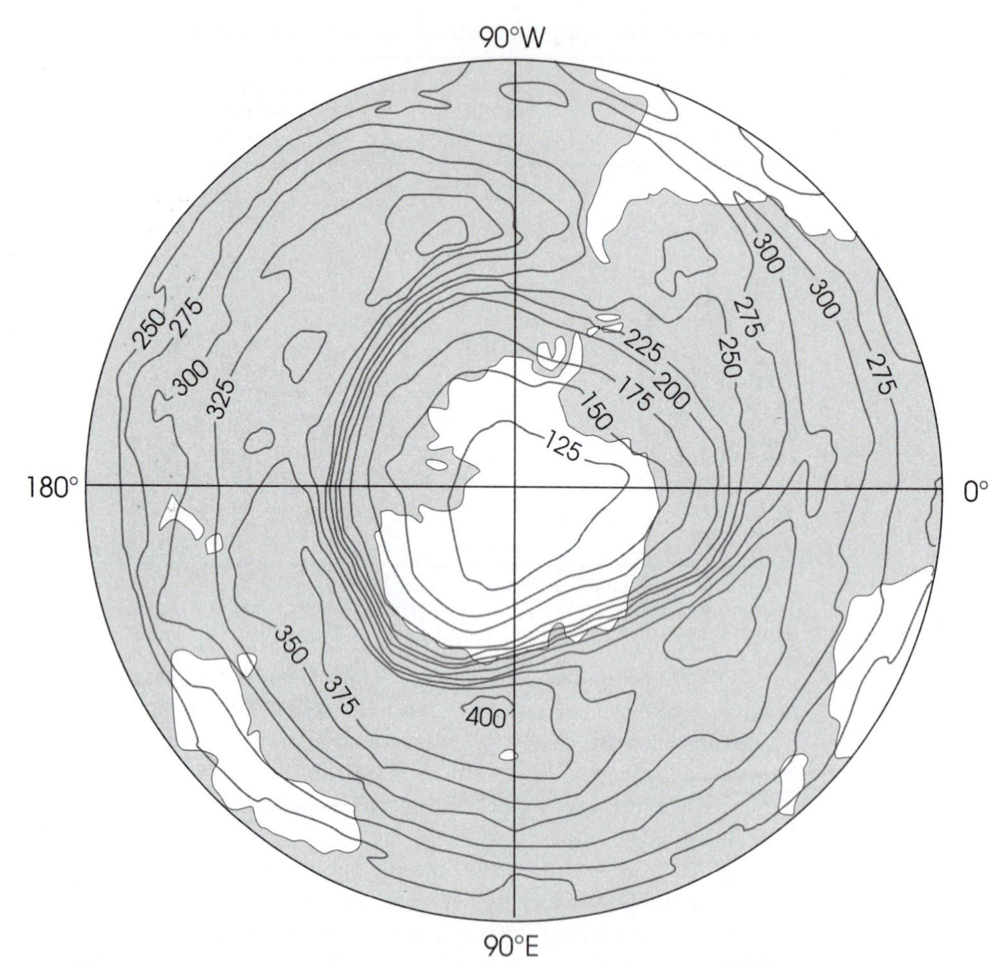


**Figure 14.3** The antarctic ozone hole over the southern hemisphere on September 28, 1992, as measured by the total ozone mapping spectrometer (TOMS) aboard the *Nimbus 7* satellite. Notice that the area of lowest ozone concentration is larger than Antarctica, and that the ozone hole is nearly centered over the south pole.

ozone concentration was only about half the surrounding region. That fall, the ozone hole, the area of maximum depletion, covered nearly half the antarctic continent.

In the antarctic a key chemical process involves molecules of chlorine monoxide (ClO) combining to form $Cl_2O_2$. When exposed to sunlight, the $Cl_2O_2$ breaks down into two chlorine atoms and an oxygen molecule. This frees the chlorine atoms to combine with ozone and repeat the process. In 1986 and 1987 ground-based measurements showed large concentrations of $Cl_2O_2$ between 17 and 23 kilometers (10 and 14 miles). This is the altitude range where most of the ozone disappears. To further confirm the chemical reactions the data showed that the chlorine monoxide concentration decreased at night and increased rapidly after sunrise.

In the late summer of 1987 scientists began an airborne antarctic ozone experiment. NASA coordinated the program sponsored by a variety of agencies from four countries. Based in Punta Arenas, Chile, the project involved flying a DC-8 and a U-2 aircraft over the antarctic to sample atmospheric chemistry. The project began in mid-August and lasted through September.

At the time of the flights, the ozone-depleted region extended from 12 kilometers up to 23 kilometers (7.4 to 14 miles). Chlorine monoxide was concentrated above 18 kilometers (11 miles) and in the zone of depleted ozone. The analysis showed concentration of chlorine monoxide of 100 to 500 times those of midlatitudes. There was also very little hydrochloric acid, a nondestructive reservoir of chlorine. There was a low concentration of nitrogen compounds which also form an inactive chlorine reservoir. These compounds include nitric oxide, nitrogen dioxide, and nitric acid. Levels were about one-tenth those of the normal atmosphere.

In the winter over Antarctica a very large mass of extremely cold, dry air keeps out warmer air surrounding the continent. This cold air gets even colder during the months when there is no sunlight. Monthly mean temperatures drop as low as −84°C (−119°F). In the extreme cold, moisture condenses into ice crystals, and nitric acid crystals form. These crystals form very high thin clouds called polar stratospheric clouds (PSCs). The cloud crystals play a very important role in the chemistry of the CFCs and ozone depletion. The nitrogen oxide crystals drop out of the stratosphere, leaving behind the chlorine and bromine compounds and the ice crystals. Each ice crystal provides a place for accelerated chemical reactions. The chemical processes are more rapid where there is a surface upon which the reaction can take place. Ice particles are good surfaces and are some 10 times as efficient as the surface of water droplets. This partially explains the speed with which the process takes place in the antarctic spring (September and October). It also explains why the process is less effective in low latitudes.

The chemical process begins when sunlight appears in the spring. The warming increases the rate of chemical reactions, and chlorine destroys ozone at a rapid rate. The depletion actually begins first near the antarctic circle, where sunlight begins to penetrate the stratosphere. It may begin here by mid-August. Spring over the south pole occurs in September and October. During this time the ozone level drops until there is no more ozone left, or the clouds evaporate. There may be a total loss of up to 60 percent of the ozone in the center of the Antarctic hole. At some altitudes it is 90 percent. Eventually, air from surrounding regions flows into the area and ozone levels recover. Polar stratospheric clouds disappear with the spring warm-up. The same process takes place elsewhere in the atmosphere but at higher altitudes and at slower rates.

There are at least two elements associated with the decrease in ozone over the antarctic continent. The obvious one is the increase in CFCs and the other is meteorologic. The meteorologic element involves changes in circulation from season to season over the region and a high-altitude change in radiation. On September 5, 1987, over an area of 3 million square kilometers (1.1 million square miles) the ozone decreased by 10 percent from previous weeks. This sharp decrease must have been due to the inflow of air low in ozone. Weather conditions also play a role in the size of the hole. The chemical processes operate over the region from 68°S to the south pole. The ozone-depleted area extends out to latitude 45°. The circulation must carry the ozone-depleted air outward from the source area. There is a temperature change taking place over the antarctic continent as well. The stratosphere over the southern hemisphere cooled by 2°C (3.6°F) between 1980 and 1985. Over the antarctic the stratosphere cooled between 2 and 4°C (3.6 and 7.2°F).

The antarctic ozone hole cannot grow continuously larger. It is restricted by the subpolar winds that circle the continent. It is in the isolated cold air inside this stream where the rapid decline takes place in the spring. Although the ozone hole is limited in size, the overall problem is not. When the antarctic air flows away in the spring, the ozone-depleted air flows into midlatitudes. When the ozone hole broke up in December 1987, large masses of ozone-depleted air moved northward. In Australia, ultraviolet radiation levels increased 14 percent above normal for December.

In the fall of 1989 the ozone content of the stratosphere over the antarctic once again plummeted (Figure 14.4). Before 1989 the worst year was 1987. The depletion of 1989 was as severe

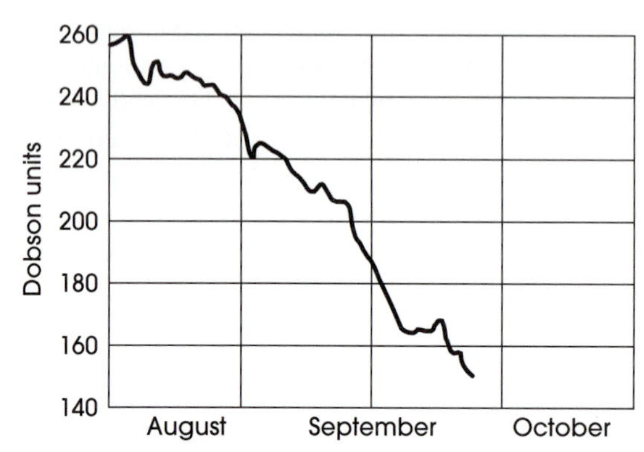


**Figure 14.4** Ozone depletion in the fall of 1989 over the antarctic continent. The sudden drop in late September and early October is due to the flood of sunlight into the stratosphere over the antarctic at that time.

as 1987. In the fall of 1989 ozone almost completely disappeared from some zones in the stratosphere. In the zone from 16 to 18 kilometers, ozone dropped 90 percent from August to the first week of October. In both 1987 and 1989 the geographical area over which there was depletion extended over an area twice the size of the antarctic continent. The average drop in ozone was almost the same in 1989 as it was in 1987. In 1987 and in 1989 the average ozone dropped 40 percent during the spring. Ozone fell sharply again in the fall of 1990. This was the first time that back to back years of ozone depletion occurred. In October 1990 there was practically no ozone at heights between 15 and 18 kilometers (9.3 and 10.8 miles). By October 4 a dramatic drop of almost 40 percent took place.

In each of the three years with severe drops in ozone, the stratosphere in the southern hemisphere midlatitudes was stable during the winter and early spring. The sunlight first reaches the south pole about mid-September. The stable antarctic atmosphere favored the development of a strong flow of upper-level winds around the antarctic over the southern ocean. This high-altitude airstream separated the cold pool of air from the warmer air of midlatitudes. This allowed the cold pool to become exceptionally cold. Temperatures within the cold pool dropped to −85°C (−121°F).

In the spring of 1991 record depletion of ozone again occurred in the Antarctic. Record lows in ozone occurred in September. By mid-November the system broke up and ozone levels recovered. The depletion usually takes place at altitudes of 12 to 22 kilometers. This is the range where most of the polar stratospheric clouds form. In 1991 ozone concentrations dropped to record lows slightly below those of the worst three previous years. On August 28 the ozone concentration was about 270 Dobson units. On September 4 it had dropped to as little as 30 to 40 Dobson units. In this year, depletion occurred at levels not attained before. Ozone reductions of nearly 50 percent took place at altitudes of 11 to 13 kilometers and 25 to 30 kilometers. The result of this depletion was a reduction of the total ozone column of 10 to 15 percent more than in past years.

The presence of volcanic aerosols offers an explanation for the unusual depletion. The eruption of Mount Hudson at 46°S is probably responsible for the depletion in the lower region. The low-level aerosol cloud arrived over the antarctic in September and the sulfuric acid particles provided the catalyst for the depletion. The temperature in this zone is too warm for polar stratospheric clouds to form to provide the catalyst.

Ozone formation in the Stratosphere

$$O_2 \xrightarrow[\text{180 – 240 microns}]{\text{Ultraviolet radiation}} O + O$$

$$O_2 + O \longrightarrow O_3$$

Ozone breakdown

$$O_3 \xrightarrow[\text{200 – 320 microns}]{\text{Ultraviolet radiation}} O_2 + O$$

Chain initiation and elimination of ozone

$$HCl + OH \longrightarrow Cl + H_2O$$
$$Cl + O_3 \longrightarrow ClO + O_2$$
$$ClO + O \longrightarrow Cl + O_2$$
$$ClO + NO \longrightarrow Cl + NO_2$$

Chain termination near the Troposphere

$$Cl + CH_3 \longrightarrow HCl + CH_3$$

**Figure 14.5** Ozone is continually being formed and removed in the stratosphere by natural processes. The addition of CFCs and other chemicals containing chlorine causes a chain reaction that removes the ozone and leaves the chlorine to remove more ozone. The process terminates when the chlorine is incorporated in more stable compounds in the troposphere.

The aerosol from the eruption of Mount Pinatubo had surrounded the antarctic vortex at high altitude by mid-August 1991. The sulfur solutions provided the catalyst for ozone depletion at heights above those where polar stratospheric clouds exist.

## *GLOBAL DECLINE IN OZONE*

Ozone depletion is less outside the antarctic because the stratospheric aerosols are less abundant and consist of liquid sulfuric acid droplets rather than ice. This difference is significant. There is no arctic ozone hole like that of the antarctic. Temperatures are warmer and there is more variable weather in the arctic, which provides less favorable conditions for the necessary chemical and circulation processes. Ozone levels in the high latitudes of the northern hemisphere have dropped 5 percent since 1971. In 1988, researchers in Thule, Greenland, measured increased concentrations of reactive chlorine compounds. These are the same compounds known to be present over the antarctic while depletion takes place. Experiments in 1989 showed the presence of the ozone hole and provided detailed measurement of the amount and extent of the depletion.

In the low and midlatitude stratosphere there is greater solar radiation during the winter months, and there is a general absence of polar stratosphere clouds. In this part of the atmosphere, the destruction of ozone results from a combination of chemical processes. Models of the atmosphere show that nitrogen oxides play a leading role in the destruction of ozone. Particular nitrogen oxides ($N_2O_5$ and $ClONO_2$) react on the surface of sulphuric acid solutions that are similar to stratospheric aerosol particles. Sulfate particles exist throughout the lower stratosphere. They form from biological and volcanic activity. The eruption of Mount Pinatubo in June 1991 injected some three times as much sulphur into the stratosphere as did El Chichon in 1982. Since there is normally an abundance of sulphuric acid particles, the increase in these particles did not increase the rate of depletion.

New evidence keeps appearing that supports a decline in the global ozone layer. In 1986, Canadian scientists detected a thinning of ozone over the arctic region. In 1988, NASA established that the global ozone layer was declining faster than expected. By 1990, NASA reported spring losses of ozone in midlatitudes of the northern hemisphere two to three times greater than before. Also in 1990, British scientists reported an accelerated rate of ozone loss over western Europe. By 1994 there was a 4 to 5 percent decline in stratospheric ozone worldwide.

## GLOBAL RESPONSE TO OZONE DEPLETION

Concern over the possible connection between CFCs and ozone loss led to a ban on the use of these compounds as aerosol propellants in the United States effective in 1987. This was part of an EPA ban on all nonessential use of CFCs. The United Nations Environment Program called a conference in Montreal, Canada, in September 1987 to discuss the possible effects of CFCs on stratospheric ozone. Representatives of more than 30 countries took part in the conference, which drafted a treaty restricting the production of CFCs.

The agreement is officially termed the Montreal Protocol. It called for freezing the domestic consumption of the chemicals at 1986 levels by July 1990 and limited production at 110 percent of 1986 levels. Included in the freeze are CFCs-11, 12, 113, 114, and 115. Nations agreeing to the protocol were to reduce consumption 20 percent by 1994, and by 50 percent of 1986 rates by 1998. Production will continue to provide a supply of the chemicals for developing countries which cannot afford to take alternative measures quickly. The protocol permits developing countries to continue production and exceed current levels if necessary for economic development. They may increase consumption up to 0.3 kilogram (0.6 pound) per capita. The protocol also freezes the consumption of the Halons 1211 and 1301, a more destructive but less prevalent class of chlorine compounds used in fire extinguishers.

### The 1989 Helsinki Conference

In May 1989, 81 countries met in Helsinki, Finland to reconsider the problem. The conference adapted a declaration calling for a complete ban on CFCs by the year 2000. They also called for a ban on Halons as soon as possible. This conference, called just two years after the Montreal Protocol was adopted, showed the seriousness with which nations considered the problem. Another meeting took place in London in June 1990. The participants tightened the schedule for halting the use of ozone-depleting chemicals. The agreement called for a total ban on CFCs, Halons, and carbon tetrachloride by the year 2000. It further called for a ban on methyl chloroform by 2005.

### The 1992 Copenhagen Conference

At the 1992 international conference in Copenhagen, Denmark, the attendees agreed to set earlier deadlines for ending the use of CFCs and other ozone-depleting chemicals. Some models show the pesticide methyl bromide to be a major contributor to ozone depletion. The 1992 agreement did not restrict the use of this pesticide.

Some countries have passed laws that are more stringent than those set by the Montreal accord. Germany passed a law that banned production of CFCs by 1995. Australia, Sweden, and Norway banned halons after 1995. Canada will stop CFC production in 1997 and stop the sale of methyl chloroform by the year 2000.

Industry responded to the pressure for action by working on substitute chemicals. In 1988 AT&T announced that it had successfully tested a substitute for CFC-11. This is the standard industry solvent used in cleaning circuit boards in the electronics industry. Called BIOACT EC-7, it was originally developed by Petroform, a Florida firm. Dupont developed a substitute for CFC-12. By 1990 the packaging industry stopped using CFC-11 and CFC-12 for making fast-food containers. It now uses a replacement compound, HCFC-22. Reductions also occurred in the use of Halons in fire extinguishers. The use of CFCs in cleaning electronic parts also slowed. More recent agreements banned the production of CFCS after 1995.

By the year 2000, major U.S. producers expect to supply nearly half the market for CFCs with hydrofluorocarbons (HFCs) and hydrochlorofluorocarbons (HCFCs). Substitute chemicals are not without their problems as well. Some of these compounds yield free chlorine. Others, such as ammonia, are flammable.

## *THE HARMFUL EFFECTS OF OZONE DEPLETION*

There are two areas of concern about the possible reduction in the ozone layer and an increase in ultraviolet radiation reaching ground level. The first problem concerns public health. The second is the role of CFCs in potential global warming. In this chapter we deal with the effect of ozone depletion on living organisms. The relationship of ozone depletion to global warming is dealt with in Chapter 19.

### Ultraviolet Radiation and Living Organisms

Stratospheric ozone filters out ultraviolet radiation in the range 280 to 320 nanometer. This is the high-energy portion of the ultraviolet radiation spectrum known as ultraviolet B (UVB). UVB radiation is also very harmful to living organisms. While the atmosphere blocks most UVB radiation, it does not block all of it. Plants did not flourish on Earth until there was enough atmosphere and ozone to block much of the UVB radiation.

All plants and animals now existing on Earth that live in sunlight have adapted to ultraviolet radiation. Plants vary widely in their tolerance of UVB. Plants that developed in climates with high-intensity sunlight show a variety of defense mechanisms for UVB. Some produce clear or nearly clear pigments that absorb UVB radiation. Marijuana plants produce protective chemicals called cannabiniods. These are also the main hallucinogenic ingredients in marijuana. In arid climates plants develop thick, shiny leaves. Cacti and olive trees are examples.

Although sunlight is essential to most life, there is a limit to how much sunlight is good. One of the effects of ozone depletion is to let more ultraviolet radiation through the atmosphere to the surface. Ultraviolet B can damage DNA. DNA is the genetic code in every living cell. Most living organisms are subject to damage by UVB radiation. Since plants cannot adjust their behavior to changing solar radiation, some are damaged by UVB radiation. The soybean is one such plant. Excessive amounts of UVB slows growth and reduces yields. Soybean yields may drop 1 percent for each 1 percent drop in ozone.

Animals and humans also have adapted to UVB radiation. Nearly 90 percent of marine species living in the surface water surrounding the Antarctic continent produce some form of chemical sunscreen. Humans manufacture melanin in the skin. This is a pigment that blocks ultraviolet radiation. A summer tan results from increased production of melanin. Persons with very fair skin do not readily manufacture melanin and sunburn very easily.

### Ultraviolet B and Human Health

Exposure to ultraviolet radiation results in aged skin, skin cancer, and a weakened immune system. The main element in the increase is the popular need for a suntan. It is ultraviolet radiation that produces tanning of the skin and also sunburn. The risk of skin cancer is much greater from overexposure, as in a sunburn, than from steady low doses. A single blistering sunburn in a person 20 to 30 years of age triples the risk of skin cancer.

### Melanoma

One form of skin cancer is melanoma. It may start in or near a mole. This involves the cells that give the skin its color and often are a mixture of black or brown, sometimes with red or blue areas. These moles continue to grow and have irregular borders. It is the least common but the most lethal form of skin cancer. The fatality rate from melanoma is about 25 percent. It is almost always fatal if it spreads to other parts of the body. Early treatment results in a survival rate of more than 80 percent. The disease is now almost epidemic in the United States. In 1974 there were about 9000 cases diagnosed. Some 27,000 new cases were expected in 1990 (American Cancer Society, 1990). This represents a tripling of the number of cases in 16 years.

The highest incidence of melanoma occurs in individuals who do not tan easily. The incidence is increasing at about 4 percent each year. There is the same rate of increase in other countries where sunbathing and tanning salons are in vogue. Younger and younger persons are diagnosed as having melanoma. When first regularly reported it was in persons aged 40 or over. By 1990 it was frequent in the age group from 20 to 40. Fatalities in 1990 consisted of about 6300 cases of melanoma and 2500 cases of other kinds of skin cancer.

### Basal and Squamous Cell Carcinoma

Other forms of skin cancer are basal cell carcinoma and squamous cell carcinoma. Basal cell carcinoma is the most common. It is a slow-growing cancer that usually begins with a small, shiny, pearly bump or nodule on the head, neck, or hands. It can bleed, crust over, then open again. It is not life threatening. Squamous cell carcinoma may start as nodules, or red patches with well-defined outlines. It typically develops on the lips, face, or on the tips of the ears. It can spread to other parts of the body and enlarge. These skin cancers can be removed by simple surgery and are rarely fatal. In 1990 more than 600,000 new cases of nonmelanoma cancer were forecast. It is likely that nearly everyone in the United States over the age of 30 has some skin damage from solar radiation. At current rates at least one in seven Americans will develop some form of skin cancer.

### Variables in the Risk of Contracting Skin Cancer

There are several variables in the risk of contracting skin cancer. The risk is about twice as high in southern United States as in northern United States. This is a result of greater intensity solar radiation in southern United States. Persons who work outdoors have nearly twice the risk as those who work indoors, resulting from greater exposure to sunshine. Persons who do not tan easily and work outdoors have about triple the average risk of contracting melanoma. A greater risk factor than ability to tan is the presence of moles on the skin. Moles on the lower leg are a particularly good predictor of risk. These are a better predictor than the number of arm moles or number of moles on the body.

Age is a factor in the risk of contracting melanoma. Persons aged 20 to 40 who received a blistering sunburn at some time in their life face a risk of melanoma five to six times greater than those that have not had a blistering burn. Melanoma risk increases with cumulative sun exposure, especially for persons over age 60. Children are particularly likely to get skin damage from UVB radiation because of their fair skin and outdoor activity. Children should avoid overexposure to sunlight because skin damaged early in life is more likely to develop into skin cancer later. By the age of 18, most people have incurred half of the total damage they will sustain over their lifetime. Since UVB may have increased in the past two decades, children are still more at risk.

There may be other health risks associated with excessive amounts of ultraviolet radiation. It may reduce the immune system. Excessive exposure may lead to increases in a variety of diseases, such as herpes simplex, leprosy, lupis, and tuberculosis. It is also linked to increased incidence of eye cataracts.

### Reducing the Risk of Skin Cancer

The risk of getting skin cancer can be reduced with reasonable care. The first rule is to avoid exposure to the midday sun. The most dangerous hours are 10:00 A.M until 2:00 P.M. local time (11:00 A.M. to 3:00 P.M. daylight savings time). There is an old saying, "only mad dogs and Englishmen go out in the noonday sun." If exposure to the sun is necessary, use a sunscreen with a rating of 15 based on ultraviolet B radiation. Ultraviolet A is also harmful to health, but not nearly as much so as ultraviolet B. Lotions with a rating of 15 provide protection from both UVA and UVB radiation. There is no evidence that sun screens with a higher rating provide additional protection. Avoid tanning parlors, as the radiation is as bad or worse than natural sunlight.

## *FUTURE DEPLETION OF OZONE*

One of the problems associated with ozone depletion is that maximum depletion may not occur until between 2010 and 2020. The antarctic ozone hole may not fill until as late as 2075. Each CFC molecule has a lifetime of up to 30 years. It takes these molecules six to eight years to rise to the stratosphere. Even if present production of CFCs and related compounds stops, there is a huge quantity of CFCs in old refrigerators, air conditioners, and foam packaging. Production has not stopped. Under present global arrangements, nearly half as much CFCs can be produced in the future as has been produced altogether since the chemicals were introduced.

Compliance with existing agreements will slow but not stop the accumulation of the chemicals in the stratosphere. Concentrations may grow to as much as 30 times the 1986 levels. Computer models show that at least an 85 percent reduction in CFCs and Halon use is needed to stabilize the level of the chemicals in the atmosphere. The chlorine already released will continue to remove ozone for at least a century. If the release of CFCs stop, there will be a lag of several decades before maximum ozone depletion takes place. Only after that time can the rate of removal begin to decline. The chlorine content may continue to rise until it reaches a level as much as six times the 1986 levels. If the release of the compounds continues unabated, a reduction in ozone of about 10 percent will take place with a possible range of 2 to 20 percent.

It is uncertain whether there is a real long-term decrease in ozone. It is possible that it is the shortness of the record and the accuracy of the data or is just a natural variation in the stratospheric gases. However, the only significant source of chlorine in the atmosphere is human industrial activity, and atmospheric chlorine doubled from 1965 to 1985. If the change is due to human activity, the effects of the CFCs are far greater than thought originally.

Decreases in ozone will result in magnified increases in ultraviolet radiation reaching the ground. Models show that a 16 percent reduction in ozone will result in a 44 percent increase in UVB radiation. A 30 percent global reduction in ozone will produce a doubling of surface UVB radiation.

Forecasts are for the incidence of all types of skin cancers to increase. The EPA forecasts that for every 1 percent decrease in the ozone layer there will be 3 percent increase in non-melanoma skin cancer. They are now forecasting an increase in skin cancer cases to rise to 12 million in the United States within 50 years. Annual deaths from skin cancer may rise to 200,000 a year in the same period.

## SUMMARY

Ultraviolet radiation from the sun reaching Earth's atmosphere creates a layer of triatomic oxygen we call ozone. This process absorbs about 98 percent of the high-energy radiation. One of the many classes of chemical compounds developed by humans are the chlorofluorocarbons. Industry uses these chemicals extensively in refrigeration equipment, aerosol sprays, and other processes. These chemicals escape into the atmosphere and rise into the stratosphere. In the stratosphere, radiation breaks the compounds down in a fashion that releases chlorine. The chlorine removes oxygen ions and reduces the amount of ozone. With a reduction of ozone, there is increased ultraviolet radiation reaching the surface. This may alter the planetary energy balance and also be detrimental to the health of plants and animals.

## BIBLIOGRAPHY

American Cancer Society. 1990. *Cancer Facts and Figures*. Atlanta: American Cancer Society.

Behrenfeld, M. and J. Chapman. 1991. Our disappearing ozone shield. *Currents*, 10:13–17.

Crutzen, P. J. 1992. Ultraviolet on the increase. *Nature*, 356:104–105.

Elkins, J. W., T. M. Thompson, T. H. Swanson, J. H. Butler, B. D. Hall, S. O. Cummings, D. A. Fisher, and A. G. Raffo. 1993. Decrease in the growth rates of atmospheric chlorofluorocarbons 11 and 12. *Nature*, 364:780–783.

Fisher, J. and R. Kalish. 1990. *Global Alert: The Ozone Pollution Crisis*. New York: Plenum.

Hofman, S. J., J. M. Oltmans, J. M. Harris, S. Solomon, and T. Deshler. 1993. Observation and possible causes of new ozone depletion in Antarctica in 1991. *Nature*, 359:283–287.

# chapter 15

# Habitat Destruction, Alien Species, and Biodiversity

***CHAPTER SUMMARY***



From the time that life appeared on earth, species have evolved and disappeared. Biological diversity (biodiversity) describes the variety of living organisms that exist in an ecosystem or on the planet. The evolution of new species contributes to the diversity of species on Earth (Figure 15.1). New variations of species are always appearing through a variety of means, including mutation. How rapidly new species develop simply is not known, but there are many estimates. The rate at which new species originate depends on how many species now exist on Earth. There is disagreement as to the number of species that are currently cataloged. It ranges from 1 to 1.4 million. Estimates of the total number range from a minimum of 5 million to a maximum of 30 million. Ten million seems to be a median estimate.

The rate at which new species evolve depends on the average life of a species. This also is not known. It is probably between 100,000 years and 1 million years. Dinosaurs were successful for 100 million years. Two current species that have persisted for a long time are cockroaches and horseshoe crabs. Cockroaches developed during the Carboniferous Period of the Paleozoic. Horseshoe crabs have been in existence for some 200 million years. The best guess range for the number of new species evolving each year is from 10 to 100. One reason for the lack of data is that it is very difficult to observe the origin of a new species. This is due primarily to the fact that

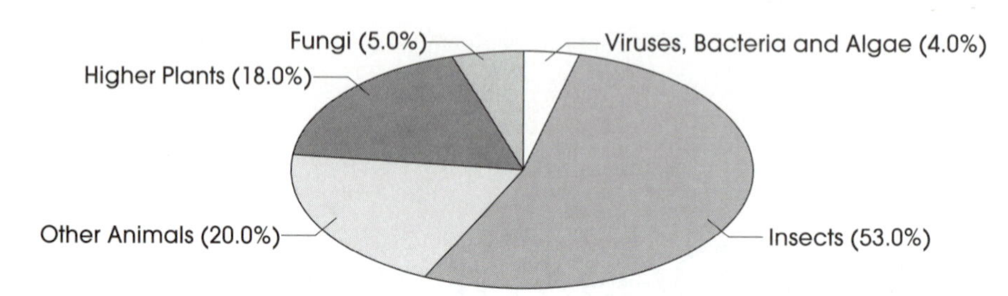


**Figure 15.1** Distribution of living species by group. The percentages are based on a total number of cataloged species of approximately 1.4 million.

most species on the planet are insects and other small invertebrates. These species are not well classified yet and often are overlooked.

Some new species of birds and mammals are discovered. In 1990 the lion tamarin monkey was discovered on a small forested island off the coast of Brazil. In the early 1990s two new species of birds were discovered. One is the Peruvian parrotlet, a flightless rail found in the Solomon Islands. The other is the striped babler, found in the mountains of the Philippines.

## BIOLOGICAL EXTINCTION

A second process that determines biodiversity is the elimination of species through disasters or failure to adapt to changing environmental conditions. Biological extinction is a natural part of the evolutionary process. Of all species that have lived on earth, probably 98 percent evolved and died away before humans appeared on earth. No species of fauna has existed for a very long period of the earth's history. Many species disappeared in mass extinctions resulting from global disasters. Some species disappeared because they became overspecialized and could not adapt to changes in the environment. Species must adapt to habitat change or die. When species become extremely specialized, habitat change often means extinction because there is too little time for species to adapt. For 30 million years primitive sponges dominated the seas. There was a large variety of species and large numbers of some individual species. They disappeared as ocean conditions changed.

Both evolution of new species and extinction of existing species take place simultaneously through time. Species diversity at any point in time depends on the rate that new species appear relative to the rate at which they are eliminated. Species diversity has grown through geologic time. At times, however, mass extinctions have occurred which greatly reduced biodiversity, as we have seen in earlier chapters.

## MODERN PERIOD OF EXTINCTION

Humans are now exterminating species at an even higher rate than did stone-age cultures. The rate of extinction has been proportional to the growth of human population. It is possible that we have started a human-induced mass extinction. If present trends continue, humans may be responsible for the elimination of half of the species on the planet before the end of the twenty-first century. This mass extinction could be more severe than any in the previous history of Earth. It may be catastrophic for the human species as well. There are several processes responsible for the increased rate of species loss. They include:

1. Habitat loss through conversion of forest and rangeland to cropland
2. Elimination of pest species
3. Overhunting for food, commercial products, and sport
4. Introduction of exotic species
5. Pollution

The International Union for Conservation of Nature and Natural Resources places the beginning of the most recent period of rapid extinction in the year 1600. This date is the earliest date for which the union believes that it has accurate descriptions of species which have become extinct. Since 1600, 36 of 4226 species of mammals have become extinct. About 9000 species of birds are known to exist. One hundred eight species of birds have become extinct since 1600. Few discoveries of new species of mammals or birds occur. The number of species of insects that lived on earth in the year 1600 will never be known.

Another 120 species of mammals and 187 species of birds are in danger of extinction. Plant and animal species are becoming extinct at a rate of 24 to 400 species a day. The number of species that have become extinct and those approaching extinction is beyond the limits of chance for a period of less than 400 years. We may be losing as many as 6000 species of plants and animals each year. This is some 10,000 times faster than new species can evolve. There is also a domino effect in extinction. It is possible that for each plant species that becomes extinct, 10 or more other species of plants, insects, and higher animals may also become extinct.

## HABITAT LOSS

In the last quarter of the twentieth century, the main factor in the elimination of species of plants and animals is the destruction of habitat. Massive elimination of Earth's natural vegetation is taking place primarily to clear land for agriculture. Agriculture eliminates natural ecosystems and creates biological systems with only one or a few dominate plants such as rice, wheat, or corn.

Deforestation is the burning, cutting, or otherwise removing virgin forests from the land. The United Nations in an assessment of tropical forests defines deforestation as the permanent depletion of the crown cover of trees to less than 10 percent of the surface area. Deforestation has, or is, occurring in nearly every country where there was a forest cover at the time the agricultural revolution began. The reasons for the elimination of forest are several:

1. Land clearing for crop cultivation
2. Land clearing for pasture
3. Tree cutting for fuelwood
4. Commercial lumbering
5. The deposition of acid rain

## TROPICAL FORESTS

Tropical rain forests exist along the equator in Asia, Africa, and South America. These forests cover about 7 percent of Earth's land area. Extensive tracts of rain forest are in the Amazon River basin, the Congo River basin, and the East Indies. Half of Earth's tropical forests were located in four countries in 1990: Brazil, Indonesia, Peru, and Zaire. Smaller tracts exist at other sites (Figure 15.2).

Tropical forests are rich in species diversity (Table 15.1). Tropical rain forests may include 50 to 80 percent of all plant and animal species on Earth. Half of Earth's bird species and 90 percent of primates inhabit the tropical rain forests. Half of the plant species that provide the basic food grains for humans originated in the tropical forests. These include rice and maize. Of an estimated 3 to 4 million species of organisms thought to exist in the tropical forests, only about 15 percent are yet classified (Figure 15.3).

### Tropical Deforestation

Destruction of the tropical rain forest is occurring rapidly. Some estimates suggest the annual loss to be an area equal in size to the state of West Virginia. In many tropical areas 1 to 2 percent of the rain forest is cut or burned each year. This is equivalent to removal of 20 to 50 hectares each minute. The area of tropical rain forest in the Ivory Coast dropped 30 percent in the 10 years from 1956 to 1966. In the decade from 1980 to 1990, some 154 million hectares of forestland was cleared for other use. Studies now show that the area of degraded and fragmented tropical forest may be greater than the area deforested. This is extremely important in terms of loss of biodiversity.

The pressure to cut these forests comes from several sources. The need for land for farming and ranching is the major reason for removal. Secondarily is the demand for tropical wood for lumber. Between 10 and 15 million hectares of land is cleared of forest in the tropics each year for agriculture and pasture. Brazil and Indonesia account for about 45 percent of global rain forest depletion. In India in the space of 10 years between 1972 and 1982, 9 million hectares of forest was cleared. This is about one-fourth of the forest that existed in 1972.

### Amazon Rain Forest

The Amazon rain forest is the largest continuous stand of forest left on the face of the earth. In 1980, the area of the remaining forest was some 5 million square kilometers, an area nearly half the size of the United States. Brazil contains about 30 percent of Earth's tropical forest. The forest covers about 40 percent of the land in Brazil. It extends into Bolivia, Columbia, French Guyana, Guyana, Peru, Surinam, and Venezuela.

For several decades the government of Brazil envisioned the economic development of this vast region. To begin development the 2500-kilometer-long Trans-Amazon highway project was begun in 1970 and completed in 1974 (Figure 15.4). The highway was funded by the World Bank. The highway, known as BR-324, runs from eastern Brazil to the western state of Rondonia. The highway is part of a scheme to develop the northwestern secton of the rain forest. Included in the plan was the clearing of 160,000 square miles of forest. The road was initially a 900-mile dirt road of poor construction. It often washed out during the rainy season. Even in its poorest form, it cut the travel time from the eastern part of the country to Rondonia from weeks to days. The road was paved in 1982 and a flood of people moved along the road, clearing land as they went. They built the highway partly to move people out of the overcrowded northeastern part of the country. The result of constructing this road and the thousands of kilometers of feeder road was disaster for the forest. A population explosion of humans and livestock followed construction of the road. Between 1966 and 1978, settlers cleared 80,000 square kilometers of forest. Most of the land was planted in grass for cattle ranching. Some estimates place the amount of original forest already cut or burned as high as 30 percent.

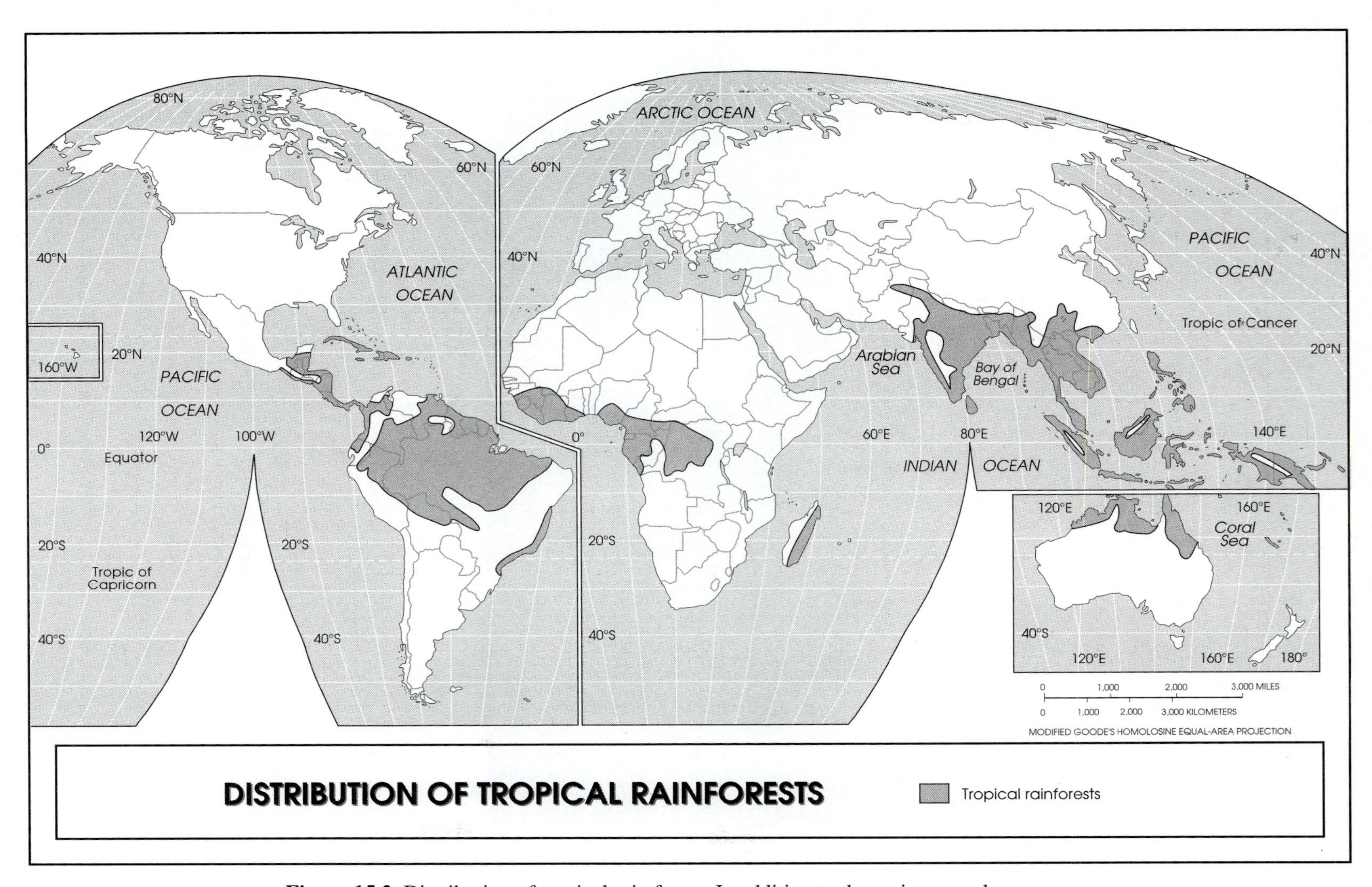


**Figure 15.2** Distribution of tropical rain forest. In addition to the main areas shown on this map, there are many small enclaves of rain forest where rainfall is high as a result of higher elevations.

**TABLE 15.1** Selected Data on Tropical Rain Forests

| | |
|---|---|
| Estimated number of species of plants and animals | 2–5 million |
| Number of species described and cataloged | 1.4 million |
| Amount of forest cleared to 1994 | 45% |
| Annual rate of loss | 1% |
| Estimated years to elimination | 20–40 |

**Figure 15.3** Natural rain forest in Nigeria.

In 1980 the government was looking to the Amazon Valley to become a source of wealth comparable to the oil of the Middle East. The government of Brazil encourages deforestation. To encourage development, the government provided incentives to foreign investors to explore for minerals and to harvest the timber. The rate of deforestation has increased rapidly since 1980. Brazilian leadership recently announced a controversial new plan to harvest an additional 40 million hectares of timber. Removing the forest at the rate existing in 1987 will eliminate 20 to 50 percent of the total forest in South America by the year 2000. At this rate of removal the entire forest in South America will be gone by 2050.

## DEFORESTATION IN MIDLATITUDES

There are rain forests in midlatitudes as well as in tropical regions. There are temperate rain forests on the west coasts of North America, South America, and Europe. They are located along coastlines where there is an onshore flow of air bringing abundant precipitation. These forests have at

**Figure 15.4** Photograph of cleared rain forest on the Amazon highway.

least 2 meters (80 inches) of precipitation each year, with most falling in the winter. In the northern hemisphere they are within a latitudinal band from 46 to 61°. On North America they exist in the states of Washington and Alaska and in the Canadian province of British Columbia. They are also found on Tasmania and New Zealand.

Midlatitude rain forests are the most productive forests on Earth. The amount of organic matter found per hectare in a temperate rain forest ranges up to twice that of the tropical forests. These forests contain the largest trees found anywhere on Earth. Some are 60 meters (200 feet) tall and 10 meters (35 feet) in diameter at the base.

Some forests in northwestern United States are defined as old-growth forests. These are stands where the trees are at least 200 years old. Some Douglas fir trees in old-growth forest are more than 500 years old. These North American forests are also home to two endangered bird species, the northern spotted owl and the marbled murrelet. Old-growth forests that were on private land have nearly all been cut. Old-growth forests on federal land will be gone by 2010 at present rates of cutting (Figure 15.5).

The original land area covered by temperate rain forests was about 31 million hectares (76 million acres). This is only about 4 percent that of the present tropical rain forest. Most of the forest in Europe is gone. In the state of Washington most is gone. In southeastern Alaska it is cut at the rate of 72 square kilometers (28 square miles) a year. The rapid rate of cutting in the Tongass National Forest of Alaska is seriously beginning to jeopardize fish and wildlife habitat. Two endangered bird species that use this dwindling forest habitat are the northern goshawk and the pine marten.

**Figure 15.5** Clearcut forest on private timberland, Tillamook County, Oregon. Douglas fir has been planted in place of western hemlock.

## U.S. Forest Service

Our national forests were established at the close of the nineteenth century to protect public forestlands from destruction by timber companies. By the time the national forests were established, much of the east and midwest had been stripped of forests. Today, the 156 national forests encompass 187 million acres, an area equal to the size of Illinois, Iowa, Michigan, Wisconsin, and Minnesota combined.

National forests offer wilderness, clean rivers, recreation, wildlife habitat, and scenic areas. The laws governing the national forests state the forests are to be managed for all users. They are to preserve wildlife habitat, protect watersheds, provide recreation, and to provide lumber. In actuality, the Forest Service is engaged primarily in selling timber rights to lumber companies. To sell the timber, the Forest Service incurs costs in surveying the parcels to be logged, in legal fees for sales contracts, and in building logging roads. In 1994 plans called for the construction of 403,000 miles of logging roads in national forests. This is more road than is in the entire Interstate Highway System.

## Sagarmantha National Park of Nepal

Nepal established a national park one of whose primary goals was to prevent deforestation of the slopes of Mt. Sagarmantha (Mt. Everest). At the time of Edmund Hillary's climb, mountain vegetation covered the slopes to well above 4500 meters. To get to the mountain was a 300-kilometer trek from Kathmandu, so there was little traffic near the mountain. In the 1960s an airfield was built near the mountain. The purpose was to aid in getting materials into the area for building schools, hospitals, and other needed facilities for the Sherpas. The added convenience of being able to fly to the base of the mountain brought a sharp increase in trekkers. People wanted to go to the mountain even if they were not going to climb to the summit. The number of visitors jumped from a few score to nearly 5000 a year. The increased number of visitors to the mountain began to deplete the supply of wood for cooking and warmth. One of the regulations of the park is that climbers and trekkers must bring in their own fuel. Whether Sagarmantha National Park can prevent the slopes of the most awe-inspiring mountain in the world from being denuded is unknown.

## *IMPACT OF DEFORESTATION*

The destruction of rain forest has global implications. On hill and mountain slopes, removal of trees often results in accelerated and extensive soil erosion. Without the tree cover, runoff takes place more rapidly and flooding occurs more often and peak discharges are higher. Deforestation releases $CO_2$ to the atmosphere, which increases global greenhouse warming. Deforestation may be the leading cause of the current reduction of biodiversity.

### Reduction in Biodiversity

The destruction of forest will eliminate from the face of Earth many species of plants and animals. Many are insect species with highly specialized requirements for close association with other species of insects and plants. In Panama, more than 1500 species of beetle live in one species of tree. Some of these insects exist in very small numbers and are restricted in their geographical area. Because of these characteristics, they are highly vulnerable to extinction. Possibly not even half of the species of insects that inhabit the forest are classified. Undoubtedly, unknown species have disappeared in recent years. Perhaps as many as 1 million species of plants and animals will become extinct by the year 2000.

### Elimination of Fish Populations

Deforestation is also hard on fish populations. In Malaysia biologists had tabulated 266 species of fish in the forest rivers. A recent four-year search found only 122 species. The majority of species found during the exploration of the region in the past century are now either extinct, extremely rare, or so localized as to have escaped detection.

### Extinction of the Iriomote Cat

An example of an animal that may become extinct due to tropical deforestation is the Iriomote cat. First discovered in the 1960s, it may soon be extinct. This unique small cat lives on a small island near Okinawa. The natural habitat of the Iriomote cat is the subtropical forest of this small island. The climate of the island is ideal for sugarcane and pineapple. The inhabitants are clearing the forest to get additional agricultural land. The habitat of the cat is disappearing in the process.

### Demise of the Dusky Seaside Sparrow

The most recent casualty to habitat destruction in the United States is the subspecies known as the dusky seaside sparrow. It was named in 1872 and was called dusky for its dark feathers. It lived in the marshes of Brevard County, Florida, near Titusville. The dusky seaside sparrow was one of four subspecies that inhabited the Atlantic coast. There are also five other subspecies on the Gulf coast. These nine subspecies are genetically very similar.

The marshes were drained to make way for agriculture and commercial development. As the marshes were drained, the population of the birds shrank. Finally, the remaining sparrows were banded so they could be monitored. When only six were left, they were captured and brought into captivity. The six remaining individuals were males. On June 16, 1987 the last one died on Discovery Island at Walt Disney World.

### Reduction of Pharmaceutical Potential

More than a fourth of all drugs sold in the United States contain ingredients found in the tropical rain forests. These are drugs that cannot yet be synthesized in the laboratory. There are a variety of antibiotics and painkillers that come from the rain forest. Others are used to treat heart disease and high blood pressure. Vincristine, a drug extracted from a tropical periwinkle, is used in the successful treatment of Hodgkin's disease. This is a form of cancer that strikes 5000 to 6000 Americans a year.

**TABLE 15.2** The last known sighting of selected species of birds

| *Species* | *Year last observed* |
| --- | --- |
| Cooper's Sandpiper | 1833 |
| Townsend's Bunting | 1833 |
| Himalayan Mountain Quail | 1868 |
| Forst Spotted Owlet | 1872 |
| Labrador Duck | 1875 |
| Passenger Pigeon | 1914 |
| Carclina Parakeet | 1914 |
| Pink-headed Duck | 1944 |
| Jerdon's Courser | 1950 |
| Dusky Seaside Sparrow | 1987 |

Note: Species are not considered extinct until 50 years after the last sighting.

## *INTRODUCTION OF EXOTIC ORGANISMS*

As organisms evolve and become more competitive, they expand their range. Natural dispersal of organisms has taken place slowly over the millennia so that some have become almost world-wide organisms. Species that are dispersed over nearly the entire earth are referred to as cosmopolitan. Species that exist over large land areas but not over the entire earth are designated continuous and may be found over a large area of a single continent or in similar environmental zones on more than one continent. Discontinuous populations are those found over large areas but not in certain regions within the main territory. Scattered populations are locally distributed in separate areal units, and endemic populations are those associated with a single geographic area.

Many species have become extinct because they were unable to compete with new species introduced into their habitat. While the spread of organisms is a natural process, humans have accentuated the process in a variety of ways. Some organisms have been carried by people inadvertently as they move about. Disease organisms, rodents, flies, lice, and rats have all moved with humans. Other plants and animals have been deliberately carried to areas where they were not found previously. Many ornamental flowering plants and trees are in this category. Also, human engineering works have aided the spread of some organisms. The development of irrigation has aided many water-related species to spread rapidly. The opening of canals has breached barriers that have stood for millions of years.

### *Anopheles* Mosquito and Malaria

The main carrier of malaria is the mosquito *Anopheles gambiae*, a species that prefers human blood. This mosquito was originally found only on the African continent, but it has slowly followed the spread of irrigation across Africa and Asia and has ultimately found its way to Australia and to the Americas (Figure 15.6). Malaria appeared in the vicinity of Rome around 200 B.C., probably being carried there by warriors from Carthage. In fact, a major epidemic broke out there in the fourth century A.D. In Italy, the carrier was not *Anopheles gambiae* but a species less prone to cohabit with human beings. The disease was associated with marshy areas, which were kept fairly well drained during Roman times. As agriculture declined following the collapse of the Roman Empire, the swamps went undrained. Such a cycle continued for 1000 years. During prosperous times malaria declined as swamps were drained. During periods of economic decay and political instability, the swamps went undrained and malaria spread. By the Middle Ages, malaria had spread across the Asian landmass. In 1929, a small colony of these mosquitos were evidently carried to the Brazilian coasts abroad a French destroyer. In the next 10 years the mosquito spread along the coast from Natal Caponga and up the Jaquaribe River valley, and local epidemics of malaria occurred. In 1938 a major epidemic that affected as many as 100,000 persons occurred in Brazil. A major eradication program eliminated the problem.

Malaria can be controlled but is still endemic to large areas of Africa and Asia. It is generally confined to farming villages where stagnant pools exist. The present containment of malaria is dependent on international agreements and controls. The control of the *Anopheles* mosquito by spraying with DDT led to some dramatic increases in human population in some areas. An extensive campaign to eliminate the insect in Ceylon after World War II led to a drastic jump in population growth rates. The percent of increase doubled in just under five years.

### Sea Lamprey

The sea lamprey, *Petromyzon marinus*, lives mainly in the North Atlantic ocean, spawning in freshwater streams. By natural migration, the species was found in Lake Ontario and some small lakes in New York. Niagara Falls proved a barrier to spread into the upper Great Lakes. In 1829, the Welland Ship Canal was completed around Niagara Falls and the sea lamprey had access to the upper lakes (Figure 15.7). The sea lamprey is a hunting predator but is best known for its characteristic as an ectoparasite. It attaches itself to fish, injects an anticoagulant and lytic fluid into the wound, and proceeds to feed on the flesh and juices until the fish dies, a period of up to a week.

The spread of the lamprey was initially rather slow. In Lake Erie, the species did not multiply rapidly, probably because of few suitable streams in which to spawn and also because of a low abundances of cold-water fish such as lake trout in the lower depths of the lake. It was reported in 1937 in Lakes Huron and Michigan and in 1946 in Lake Superior. The sea lamprey is

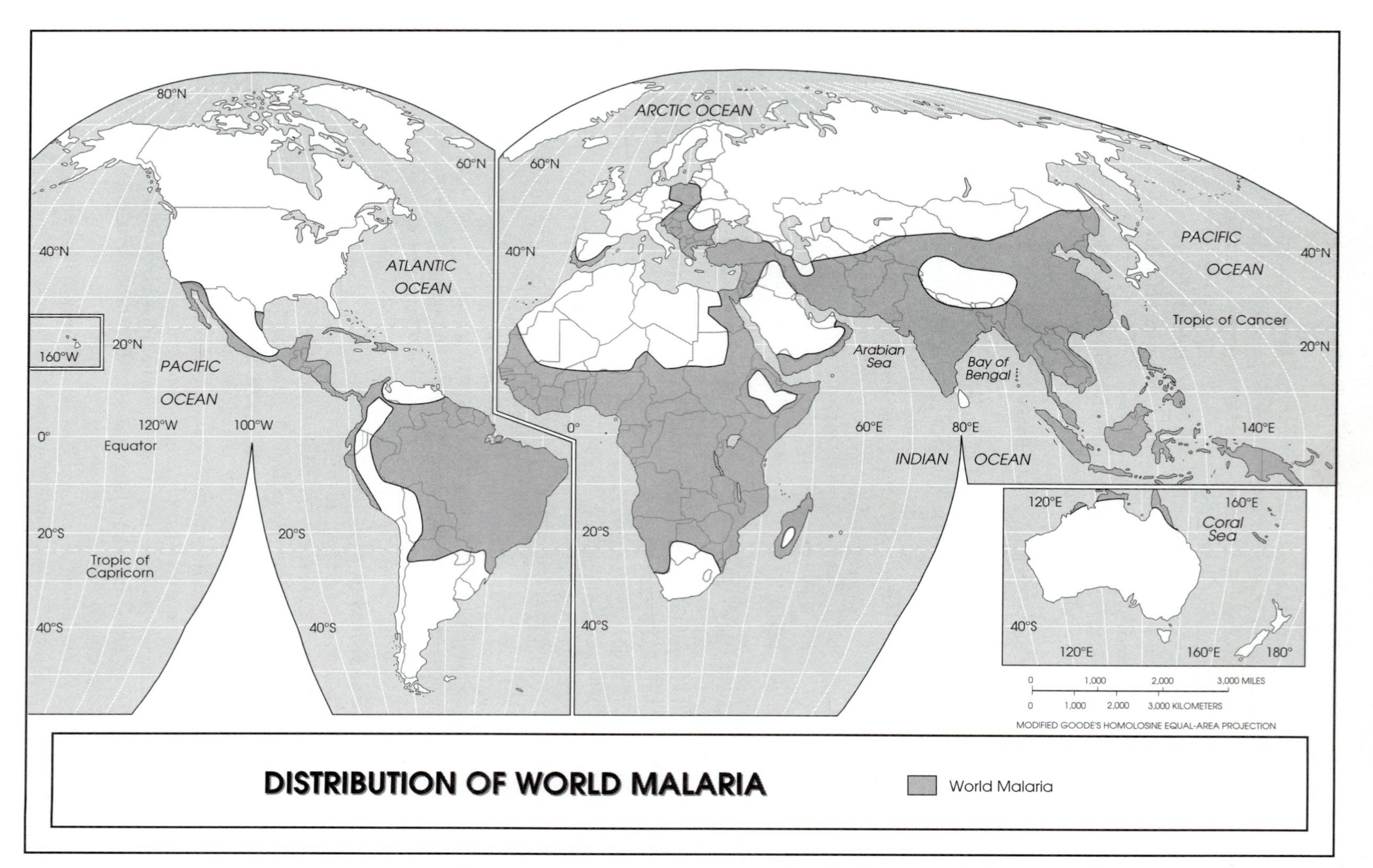


**Figure 15.6** Distribution of malaria in 1982.

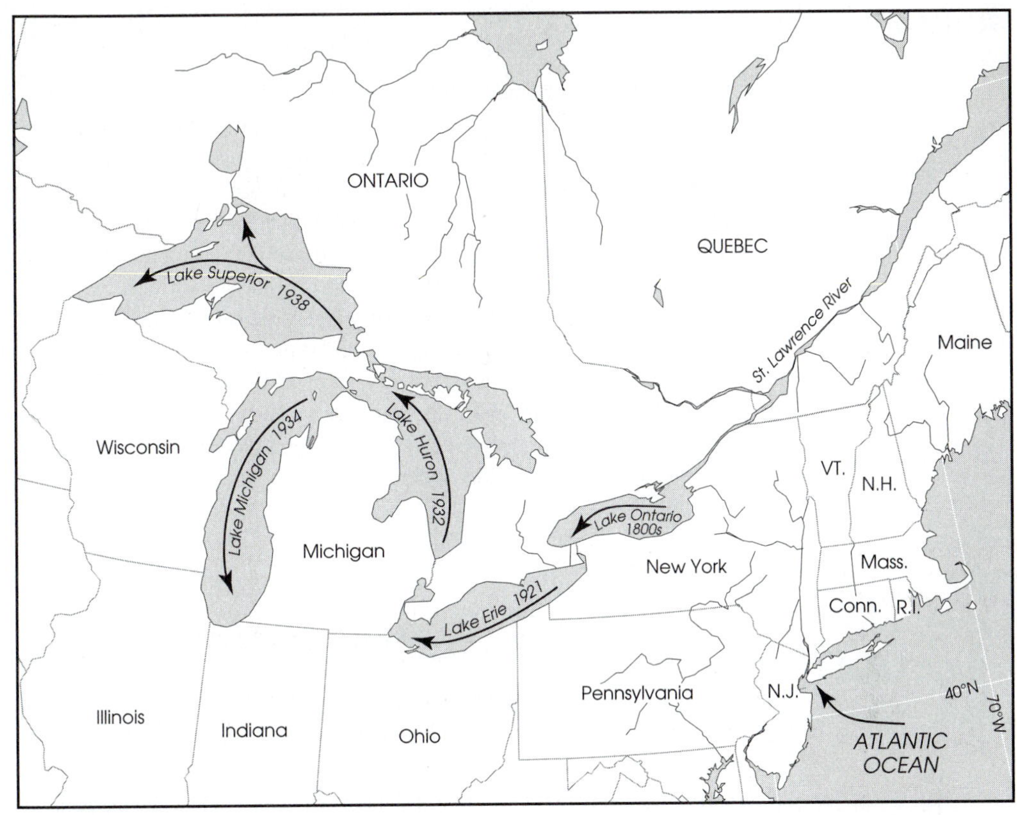


**Figure 15.7** Progress of spread of the sea lamprey through the Great Lakes. Efforts to control the population of the species are extensive. The map shows those streams where control is difficult. (From O. S. Owen and D. D. Chiras) *Natural Resource Conservation*, Prentice-Hall, Englewood Cliffs, N.J., 1995.

responsible for a whole series of events that brought chaos to fish populations in the upper Great Lakes. The lake trout were the preferred victims of the sea lamprey, as these fish are abundant in the colder subsurface strata in which the sea lamprey thrives. Other species were also affected, such as burbot (*Lota lota*). Production of lake trout in Lake Michigan in 1943 is estimated to have been 6,860,000 pounds, with essentially no loss due to the sea lamprey. In 1953, production of lake trout had dropped below 500 pounds, a level at which the species was unable to survive. Estimated destruction of fish by sea lamprey in the peak year of 1951 was some 11,744,000 pounds. The initial decline cannot be attributed to the sea lamprey alone because extremely heavy fishing was going on at the same time. The early war years increased substantially the demand for lake fish. As lake trout declined, both the fishing fleets and the lamprey turned to the chub as targets. The larger species of chub (*Leucicthys johannae* and *L. nigripinnis*) became extinct and the smaller species the bloater (*L. hozi*) and the alewife, became the major commercial species. The alewife made a major explosion in population. Experimental trawling in 1954 produced no alewives, and in 1960, 45,000 were taken in similar trawling. These were not taken for human food but for animal food and fishmeal. The smaller chubs had been the major food supply of the lake trout.

## SUMMARY

Evolution provides our planet with new species each year. This process began in the Precambrian era and continues today. No species that evolved on the planet has persisted throughout the history of the planet. Almost all perished for one reason or another. There were five major mass extinctions to the beginning of the Cenozoic Era. We are now witnessing another mass extinction. Whether it will rank with the major biological disasters of planetary history remains to be seen. In terms of the loss of terrestrial species it will probably rank among the worst.

The current mass extinction is associated with the growth of the human population. Species of land plants and animals are disappearing at a rapid rate. Two reasons for the demise of species are the destruction of habitat and the introduction of exotic species. Removal of the forests, particularly the tropical forests, is a primary form of habitat elimination. Humans transport plants and animals from one part of Earth's surface to another, usually without thought to the consequences of introducing a new species into an area where it did not exist previously. The introduced species often forces other species out. Many species have disappeared since 1600 due to these two human activities.

## BIBLIOGRAPHY

Bird, C. 1991. Medicines from the rainforest. *New Scientist*, 1782:34–39.

Brown, N., and M. Press. 1992. Logging rainforests the natural way? *New Scientist*, 1783:25–29.

Caufield, C. 1991. *In the Rainforest*. Chicago: University of Chicago Press.

Frisvold, G. B., and P. Condon. 1994. Biodiversity, conservation, and biotechnology development agreements. *Contemporary Economic Policy*, 12:1–9.

Huston, M. A. 1994. *Biological Diversity*. Cambridge: Cambridge University Press.

Sisk, T. D. and others. 1994. Identifying extinction. *Bioscience*, 44:592–604.

Tanner, J. E., and T. Hughes. 1994. Species coexistence, keystone species, and succession: a sensitivity analysis. *Ecology*, 75:2204–2219.

Wilson, O. E. 1992. *The Diversity of Life*. Cambridge, Mass.: The Belknap Press of Harvard University.

chapter 16

# Effects of Pesticides, Hunting, and Other Human Activities on Biodiversity

### *CHAPTER SUMMARY*



Monoculture is the practice of raising a single crop on large amounts of land. It has created simplified ecosystems in which some insects have become pests. There are now at least 3 million species. Of all insect species only about 1 percent, or 3000, are agricultural pests or carriers of human and animal diseases. Some of these, however, are so destructive of human and animal life, or of domestic plants, that we must destroy them. If we do not eliminate them, they must be reduced in numbers if the human species is to survive using present sources of food.

## PESTICIDES AND THE BIOSPHERE

One of the products of the modern industrial age is the class of chemicals known as pesticides. Pesticides have been used for more than 2 centuries. They have become widely used only in the past three decades. There are now nearly 1000 active chemicals used in herbicides and pesticides in the United States. These are ingredients in more than 90,000 different compounds. They enter the environment as sprays, dusts, and pellets.

One class of insecticides is a group of synthetic organic compounds referred to as chlorinated hydrocarbons. Synthetic organic insecticides are now the most efficient means of controlling insect pests. The chlorinated hydrocarbons have as their major ingredients chlorine, hydrogen, and carbon. There are three classes of synthetic hydrocarbon compounds now in use as pesticides: organochlorine, organophosphorus, and carbamate compounds. Among the organochlorine compounds are DDT, methoxychlor, aldrin, dieldrin, endrin, isodrin, telodrin, and heptachlor. The organophosphorus compounds differ in containing sulfur and phosphorus. Parathion and malathion are in this group. The carbamates contain nitrogen and include carbaryl (Sevine and Baygon).

Pesticides vary widely in their effectiveness. There are differences in response to a given chemical among species, sexes, age groups, and individuals. The chlorinated hydrocarbons are all nerve poisons. Basically, if an organism has nerves, the chlorinated hydrocarbons can kill it. As a result of this property, these compounds are poisonous throughout the range of organisms in the animal kingdom from invertebrates up to mammals. Because they are broad-spectrum poisons, they have earned the term *biocides.*

Low concentrations produce instability in the organism, and increased concentration may produce convulsions and affect breeding. Virtually all organisms, including humans, store residues. The impact of sublethal concentrations is a very important problem for organism. Even minute quantities affect the behavior of some varieties of fish.

Some of the chlorinated hydrocarbons are very stable in nature having a half-life of up to several years (Figure 16.1). Two of the classes, organophosphorus and the carbamates, usually break down rather rapidly compared to organochlorine compounds. The carbamates break down

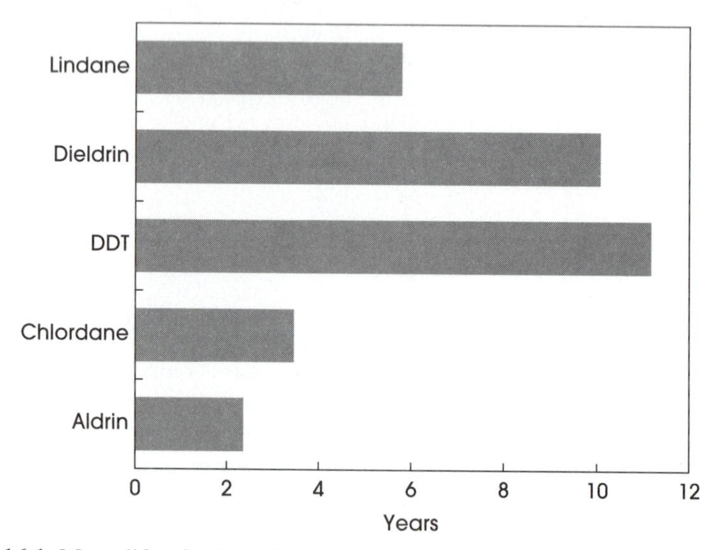


**Figure 16.1** Mean life of selected agricultural pesticides. DDT has the longest life and is still widely used in developing countries.

into nontoxic compounds in from 1 to 10 weeks. Such is not the case with DDT and the other organochlorine compounds. The persistent nature of these compounds makes them valuable in pest control. It also makes them dangerous for all of Earth's living organisms.

## *DDT*

DDT (dichlorodiphenyltrichloroethane) is the most commonly used and most understood of these compounds. This is partially because it was the first introduced on a wide scale. Its residues are the most widespread in the environment. A German chemist, Othmar Zeidler, synthesized it in 1874. It was not until 1939 that a Swiss entomologist, Paul Mueller, discovered the insecticide properties of the compound. In 1942, about 3 kilograms came to the United States for testing. The military found it to be very effective as a delousing agent and began using it for a variety of insect control purposes. After World War II, the compound was widely applied for insect control.

Production in the United States jumped from 4400 metric tons in 1944 to a peak of 81,300 metric tons in 1963. In 1948, the Nobel Prize in Chemistry went to Paul Mueller for his discovery of the insecticide properties of DDT. Following the introduction of DDT came a succession of related compounds. The organochlorine compounds do not react readily with water and are broken down very slowly by metabolism. Metabolism breaks DDT down into DDE and DDD. These residues are also toxic, and DDE is the most widespread residue of DDT.

DDT is not very soluble in water but is quite soluble in organic solvents and fatty substances; therefore, it moves from water into organic compounds. Thus the pesticide residues move not only in the air, in suspension in water, but also in living organisms as well. Bioaccumulation is the storage of chemicals by repeated eating of contaminated food. Chlorinated hydrocarbons that collect in body tissues include polychlorinated biphenyls (PCBs), DDT, and DDE. Water fleas can concentrate DDT from 0.5 parts per billion to 500 parts per billion. This is a concentration factor of 1000 times. Oysters can concentrate DDT from 1 part per billion to 700 parts per billion, a concentration of 700 times. About 80 percent of the chlorinated hydrocarbons found in marine organisms is DDE. Inorganic particles or microorganisms such as diatoms readily absorb DDE. Biomagnification occurs rapidly when larger animals eat smaller ones containing chlorinated hydrocarbons. The reason is that the predator accumulates the residues and the concentration goes up rapidly. Small marine organisms feed on a steady diet of pesticides and collect the toxins.

It is the stability of the organochlorine compounds and their residues that is responsible for their wide distribution over the surface of Earth. In some cases less than half of the spray released from aircraft reaches the ground in the intended area. The rest may rise high into the air and disperse through the atmosphere. Particulate matter in the atmosphere absorbs the pesticide droplets. It then falls out with rain. Rain occurs with a residue concentration of as high as 0.34 part per billion.

Pesticides get directly into the surface water through spraying for aquatic insects. It also gets into water accidentally from spray drifting from overland applications. Insecticides applied to soil have a long enough life to wash out as suspended droplets or absorbed in sediment. The final sink of a large proportion of pesticides is the sea.

### Biological Magnification

Once these insecticides get into the food chain, a process of concentration begins called biological magnification. The food that is taken in by the organism is oxidized and wastes given off. Most of the pesticide residue is retained. The degree of concentration may be anywhere from 2 to 1000 at each level of the food chain. As the residues flow up the food chain from one trophic level to the next, they accumulate. This magnification results in concentrations of up to 1 million times original levels in some carnivorous birds. In species at the higher trophic levels, accumulation occurs both from direct intake of pesticides and by absorbing the residues stored in the food they eat. Organisms may become contaminated from an environment that appears free of original pesticides.

In the 1960s, DDT was sprayed in the Carmans River Marsh in Long Island for control of mosquitos. Concentrations of DDT and its residues amounted to 2 kilograms per hectare. Concentrations of DDT in offshore waters were 0.00005 part per million. Study of the samples taken in Carmans River marsh showed more residue in larger individuals and higher trophic levels than in smaller organisms at lower trophic levels. Organisms at lower trophic levels contained high proportions of DDT and lower levels of DDE and DDD. At the higher trophic levels, the ratio was reversed. This results from breakdown of DDT in the magnification process. In many cases concentrations were nearly as high as those found in dead organisms. There were reduced populations of shrimp, amphipods, summer flounder, blue crab, spring peeper, and other species. Fish and birds showed concentrations of more than one-tenth that of fatal dosage.

Carnivorous birds have been particularly vulnerable to pesticides and their residues. In both Europe and North America changes in reproductive success coincide with exposure to pesticide residues. Residue concentrations develop that are from 10 to 100 times as high as those of the fish upon which they feed. Reproductive failure results mainly from death of chicks around the time of hatching. Eggshells of affected birds have a lower calcium content than eggs from unaffected birds. The result is that the eggs are much more susceptible to breakage. DDT also affects development of reproductive ability in birds. Nervousness and aggressiveness are also characteristic of affected birds.

The island of Bermuda was the nesting place for a variety of oceanic birds when European settlers arrived in 1609. Perhaps the most abundant was the Bermuda petrel (*Pterodrome cahous*). The bird came very close to extinction within a score of years as a consequence of hunting by humans and domestic pigs. However, after 1900 there were several sightings, and in 1967 twenty-two pairs nested on an islet near Bermuda. The small remaining population of birds was in danger of becoming extinct due to accumulation of pesticide residues.

The Bermuda petrel visits land only to breed and breeds only on Bermuda and its surrounding small islands. DDT was not used on these islands insofar as is known. What is more, being pelagic, the bird feeds in the sea and not on land. DDT residues in the birds can only result from the presence of the poisons in plankton far out to sea. Concentrations in unhatched eggs and dead chicks showed residues of 6.44 parts per million. From 1958 to 1967 the birth of live chicks decreased more than 3 percent per year. Had the decline continued, extinction would be a certainty before the end of the twentieth century. Other birds severely affected by DDT include the osprey (*Pondion haliaetus*), American woodcocks (*Phibohela minor*), peregrine falcon (*Falco peregrinus*), golden eagle (*Aquila chrysaetos*), bald eagle (*Halia eetus leucocephalus*), herring gulls (*Larus argentatus*), and sparrow hawk (*Accipiter nisus*).

Single applications of DDT for mosquito control have caused drastic reductions in populations of crayfish, shrimp, fiddler crabs, and fish. Spraying $\frac{1}{2}$ kilogram per hectare in New Brunswick, Canada produced high mortality in salmon. Applications of 1 to 2 kilograms per hectare have major effects on both fish and birds as well as all organisms below them in the food chain.

In the ocean, virtually all organisms become contaminated and the toxic effects are significant at all levels from the smallest green plants on up the food chain. DDT concentrations of a few parts per billion can decrease photosynthesis in some phytoplankton. Single-celled algae are responsible for more than half of the earth's photosynthesis. Destruction of these marine plants would have far-reaching effects. Although pesticide levels to which these organisms are subject are very low, mortality is high even to very low concentrations.

### Insect Resistance to Pesticides

Pesticides have acted to produce new strains of insects resistant to pesticides. The number of resistant species reached 224 by 1967. Since the length of one generation of insects is short compared to vertebrates, insect resistance develops very rapidly. When insects become resistant, it forces either heavier doses of pesticide or a switch to a different form of pesticide. Using continually heavier doses means that excessively higher concentrations are passed on to predators.

## *ILLEGAL HUNTING*

Illegal hunting for animal products is widespread. Prices for rare animal products have skyrocketed, making hunting some species highly lucrative even if the hunting is illegal. Illegal hunting of some species is increasing despite enforcement. A high demand for ivory and rhino horn ensures the hunting of elephants and rhinoceros.

### Elephant Ivory

In 1970 there were about 5 million elephants in Africa. The price of ivory climbed from $5 per kilogram in 1970 to $75 in 1982. Even a medium-sized elephant carrying 50 kilograms of ivory was worth more than $3000, a large animal, at least twice that. Poachers kill as many as 40,000 animals annually for their ivory (Figure 16.2). Most are taken by commercial poachers in countries where hunting is illegal. In Kenya poachers killed nearly two-thirds of the 150,000 elephant population in less than 10 years. The United States supported this illegal trade in ivory by massive imports. In 1980 the United States imported more than $8 million worth of ivory.

By 1989 there were only about 600,000 animals in the wild. The primary international law protecting endangered species is the Convention on International Trade in Endangered Species (CITES). The convention provides for permits for both exporting and importing threatened and endangered species and any products made from them. In 1990, an international ban on the trading of ivory was approved by members of CITES. If the international market for ivory is broken up, there may be hope the remaining population will have a chance to survive.

### Rhino Horn

A significant demand for rhino horn exists from two longstanding uses. Some Asians believe that rhino horn has medicinal value and power as an aphrodisiac. There is no known basis for either claim, but this matters little. So strong is the demand for powdered rhino horn that the street price had reached $24 per gram in 1982 (Figure 16.3).

An additional market for rhino horn exists in the Middle East. In North Yemen adult males often carry a dagger as a sign of manhood and status. Top-of-the-line daggers have rhino horn handles and scabbards with gold and silver decorations. Traditionally, few men could afford these, but the wealth of oil changed the economics of the region. Between 1970 and 1980 North Yemen imported the horn of at least 8000 rhinos. The impact of this increased demand has been decimation of the African rhino population. Of an estimated total population of 100,000 in all of Africa in 1970,

**Figure 16.2** African elephants are being actively poached over much of the continent.

**Figure 16.3** The African black rhino is more endangered than the African elephant. The rhino is more solitary than the elephant. Numbers have declined to the point that the species may become extinct.

no more than 20 percent remained in 1982. The actual number has decreased since then. It is unlikely that the population will ever recover even if poaching stops. Females produce only one offspring every four years, a fatally low rate of reproduction compared to the rate of poaching.

### North American Deer

Some wildlife experts estimate that the number of deer killed illegally by poachers in the United States is as large or larger than the number taken under legal limits. In Michigan alone, the number of deer poached may reach 300,000. The deer are not endangered. For this reason hunters believe that there is no reason not to kill them illegally.

## *WAR AND WILDLIFE*

The breakdown of political structure in the developing countries often carries with it sudden and massive destruction of wildlife. Loss occurs in protected areas such as national parks and game preserves, and in nonprotected areas. In 1971, Idi Amin gained control of Uganda. This spelled massive slaughter of game animals. Army units and others raided Rwezori National Park. They used machine guns, rockets, and artillery to see how many animals they could destroy. When the Tanzanian Army drove Idi Amin from power, the victors continued the destruction. In 1971 there were 300 elephants in the preserve. By 1981 there were only 160 left. Eighty percent of the water buffalo and more than 70 percent of the hippopotamus disappeared. In all, probably 75 percent of the populations of large animals perished in that 10-year period.

## *PREDATOR CONTROL*

The practice of predator control has reduced some species to near extinction. Governments have historically enacted control programs when humans and their livestock have moved into areas occupied by predators. In North America the wolf is one of the predators selected for elimination. As Europeans spread across the continent, domesticated farm animals replaced wild animals. Deforestation reduced the woodland habitat of wolves. Inevitably, wolves took some domesticated animals for food.

On November 9, 1630, the state of Massachusetts passed the first law offering a bounty on wolves. In 1931 the U.S. Congress passed legislation that called for the decimation of wolves, coyotes, mountain lions, bobcats, and a variety of other animals. As a result of bounty and sport hunting, populations of wolves dropped in both North America and Asia. Through systematic hunting and removal of habitat, wolves are all but gone from Mexico and the lower 48 states. Globally, there are some 90,000 wolves left. The majority (60,000) of these are in the former Soviet Union and the remainder in North America.

## *VULNERABILITY OF ISLAND ECOSYSTEMS*

Island ecosystems were very hard hit when first occupied by humans. Paleontologists have found remains of recently extinct birds on almost every island studied. Some 2000 species of seabirds may have become extinct in prehistoric times. These 2000 species represent about one-fifth of all species of birds inhabiting Earth 50,000 years ago.

### Extinction of Species on New Zealand

New Zealand and its associated smaller islands encompass an area of about 266,000 square kilometers. There are no native terrestrial animals in the islands. The higher forms of fauna consisted of a variety of large flightless grazing birds, the moas. The largest of these giant birds (*Dinornis maximus*) was more than 3 meters in height, and there were at least 27 different species. Polynesians reached the islands around the year 950. No species of moa became extinct prior to the arrival of the Polynesians. By the time the Maori people arrived in 1350, most of the moa were extinct. Bones from 22 of the extinct moa species occur in association with human artifacts. This shows the species still existed when humans arrived.

In addition to the moas, another 150 varieties of birds existed on the islands. In the 819-year period from A.D. 950 to 1769, some 36 species of birds have become extinct. They included a great eagle, flightless rails, a great swan, and two flightless geese. Nine of these species have become extinct since the year 1600. A dozen or more of the remaining species are on the endangered list. The introduction of some 35 exotic species is providing additional competition for native birds.

### Historic Example of the Great Auk

On May 21, 1534, the French explorer Jacques Cartier stopped at Funk Island, off the coast of Newfoundland. The reason for the stop was to allow his crew to feed on fresh meat. In this case it was the meat of the great auk. The great auk was a large flightless seabird that was easy to capture. In 1785 feather collectors visited the island. They killed many of the birds to use the feathers for pillows and mattresses. So popular were these feathers that the great auk was completely gone from the island by 1841.

The great auk living on other islands were doomed as well. The last one on the Orkney Islands north of Scotland was killed in 1812. They disappeared quickly on other islands near Scotland and Ireland. On June 3, 1844, two fishermen killed the last breeding pair at Eldey Rock off the coast of Iceland.

### Island Invasions by Alien Species

Island floras and faunas are especially vulnerable to invasion by exotic species. Most oceanic islands are either active volcanic peaks or remnants of extinct or dormant volcanic cones. Others, such as atolls, are coral islands built on top of submerged volcanic peaks. Since these islands have been built from the seafloor, they have never been connected with continental landmasses. To colonize them, plants and animals have had to cross open ocean, in some cases thousands of miles. As might be expected, relatively few species reach the more remote islands. The farther an island is from a continental landmass and the younger the island, the fewer species there are present.

Colonization can take place relatively rapidly, however, as was demonstrated on Krakatoa following the destruction of the island in 1883. Part of the reason for the rapid colonization is the presence of many nearby islands from which species could readily migrate.

The lack of major predators on some islands permitted the growth of very large populations of ground-nesting seabirds and the evolution of flightless land birds such as the flightless rail (*Porphyriornis nesiotis*) on Tristan da Cunha in the South Atlantic. On the less complex and oddly structured ecosystems of some islands, the introduction of alien species can cause catastrophic changes. The introduction of predators such as cats and dogs has caused the extermination of some rare flightless birds and tremendous reductions in populations of ground nesting birds. Introduction of rabbits, goats, sheep, pigs, and cattle have eliminated many forms of native vegetation that evolved in an environment where grazing animals were absent.

### Alien Species on the Island of Tristan da Cunha

Expeditions and shipwrecks deliberately and inadvertently introduced alien species. Tristan da Cunha, an island in the South Atlantic Ocean, was covered with fairly dense vegetation and sustained large colonies of ground-nesting petrels and shearwaters and substantial numbers of albatrosses and flightless rails. The island was discovered in 1506 and was settled in 1810. Rats and cats were brought to the island. Rats are believed to have come ashore from a shipwreck in 1882, and cats were apparently brought with the first group of settlers and soon were running wild. These two animals were largely responsible for extermination of the flightless rail and a great reduction in petrel colonies. In addition to cats and rats, goats, pigs, dogs, cattle, sheep, poultry, mice, and donkeys were introduced to the islands. At least six native animals are now extinct on the island and a number of others remain but in very small numbers. A similar situation exists with respect to plants. Over 50 percent of the flowering plants now found on the island are introduced specis.

Island species have in general suffered more than continental species. Of the native plant and animal species found on the Hawaiian Islands, some 10 percent are believed extinct and another 36 percent are in danger. One hundred forty-nine of some 160 birds that have become extinct since the year 1600 have inhabited islands. Of the 160 or so species of birds that have become extinct since the year 1600, 149 of them have inhabited islands.

## *THE U.S. ENDANGERED SPECIES ACT*

The Endangered Species Act was originally passed in 1973. The purpose of the act was to manage, research, and monitor populations of animals and plants whose continued existence is in doubt. The act specifies that plants and animals in danger of extinction be identified. It creates regulations governing endangered species and their habitats. There are two categories of species under the act. Endangered species are those in imminent danger of becoming extinct. Threatened species are those in danger of becoming endangered if no action is taken.

The act prohibits the federal government from undertaking projects that would further harm endangered or threatened species. It does not apply to development projects on private land but does prohibit injuring or killing threatened species on private land. In 1994 there were 418 animal species and 404 plant species listed in the two classes. Since 1973 when the act was passed, 17 species have been moved from the endangered to threatened class, and four have been removed from the threatened class. The four species removed from the threatened list include four birds on the island of Palau, a U.S. territory, and a plant in Utah. Other species have been removed from the list because of additional findings.

Hearings on reauthorization of the act began in 1994. There was a heated debate over the act. Opponents point to the controversy over the spotted owl of the Pacific Northwest. The spotted owl nests in dead trees or broken trees in old-growth forests. Without the forest in its present form the spotted owl cannot survive. The spotted owl became the rallying point upon which to try to save the remaining old-growth forests. Efforts were made to have the spotted owl listed as an endangered species. Lumber interests were adamant against the efforts to preserve the habitat of the spotted owl because it removed some good timberland from sale by the Forest Service. Under the Endangered Species Act, the government cannot take action on federal lands that would threaten the survival of the owl. Such action would prohibit logging in the remaining old-growth forests. In June 1990, the U.S. Fish and Wildlife Service formally declared the spotted owl a threatened species. The government then agreed to set aside areas of old-growth forest as conservation areas for the spotted owl.

The snail darter (*Percina tanasi*), the Alabama sturgeon, and the Bruneau snail in the hot springs of Idaho are other species that started major controversy. The snail darter was an early case which drew heated debate and ended up weakening the Endangered Species Act. The fish were found in an area that was to be flooded by the resrvoir behind a proposed new dam. It was believed to exist only in this area, and the reservoir would destroy its habitat. The dam was finally built despite the threat to the snail darter. The fish was found in other areas, so was not seriously threatened by the construction of the dam.

The *Endangered Species Act* will be funded for the duration. About two-thirds of the funds will go to the Fish and Wildlife Service. The rest will go to other agencies involved in wildlife protection. There have been successes under the act. They include the bald eagle and the arctic perigrine falcon. Both were on the brink of extinction in 1973. The bald eagle is now listed as threatened in five states and endangered in 43 of the lower 48.

## *SUMMARY*

There have been two periods of exceptionally rapid extinction of animals during the time that humans have inhabited Earth. The first was toward the end of the Pleistocene Epoch. It was about this time that group hunting became efficient enough that large animals such as mammoths could be hunted. It was also a time when many species were under natural stress from the changing global climate. A combination of the two factors resulted in the extinction of many large animals.

The second period of above-normal rates of extinction is the modern period since 1600. Many factors contribute to the current high rate of extinction. The primary factor is the elimination of habitat. The primary reason of loss of habitat is the expansion of agricultural land. The simplification of ecosytems for agriculture eliminates habitat for both plants and animals. The modern period of extinction may become one of the major mass extinctions experienced on Earth.

The use of pesticides has taken a heavy toll on populations of birds and marine organisms. It has had limited effect on the insect populations it was designed to control. Pesticides continue in use, as do accumulation of residues in living organisms.

## BIBLIOGRAPHY

CHADWICK, D. H. 1992. *The Fate of the Elephant*. San Francisco: Sierra Club Books.

NORTON, B. 1991. *The African Elephant: Last Days of Eden*. Stillwater, Minn.: Voyageur Press.

SPEART, J. 1994. The rhino chainsaw massacre. *Earth Journal*, 49:26–36.

WARD, G. C. 1992. India's wildlife dilemma. *National Geographic*, 181:2–29.

WOLKOMIR, R. and J. WOLKOMIR. 1992. A history of animal decimation. *International Wildlife*, 22:6–12.

# part III

# FUTURE GLOBAL CHANGE

Two extremely important questions that face humans are; what will the global environment be like in the future? and, what will the relationship of the human species be to that environment?

Planet Earth has always been changing. From the moment when debris began to accumulate in a mass that was to become Earth, change has been a continuous process. Earth will continue to change in the future. What is not known is how it will change. Forecasting the future is easy. Forecasting it correctly is much more difficult, if not impossible.

One of the goals of science is to be able to forecast. Future global change includes natural changes that will take place without the influence of human species. The Law of Uniformitarianism states that the natural processes operating in the environment today operated in the past. It is appropriate to conclude that these same processes will continue to operate in the future. If the planet is in a steady-state condition, then the mean planetary environment should be the same in the future as now.

Future global change must also include human-induced changes. The addition of the human element makes accurate forecasting of global change even more difficult. Accurate forecasting of human activities very far in advance has proved to be next to impossible in the recent past. This is because there are many variables related to human activity that cannot be forecast accurately. Some of these are:

a. rate of growth of the human population
b. rate of economic development

c. rate of consumption of fossil fuels
d. changes in technology
e. changes in human behavior

However, it is necessary to make forecasts of human activities, and to make them as accurately as possible. There are two aspects to these forecasts. One is the direction of change, and the other is the size of the change. Greater reliability can be placed on the direction of change, and less reliability on the amount of change. For example, present trends in industrial and agricultural activity are expected to continue in the immediate future. How far into the future they can continue is unknown.

One of the critical elements in future global change is that of Earth's climate. Forecasting natural variation in climate is very difficult. This is because there are so many forces that influence climate. Some are internal to the planetary system and others external. Before the human impact on climate can be forecast it is necessary to determine how the changes in human activities are going to vary through time, and from place to place. Because climate is such a key element in determining the global food supply, it is extremely important to develop means of making the best forecasts possible.

The remaining chapters in this book deal with some of the problems of assessing the future global environment.

chapter 17

# Forecasts and Predictions

**CHAPTER SUMMARY**



Although the future is extremely important to contemporary society, it is probably no more so to those living now than it was to people at any other time in history. The desire to know the future is deeply ingrained in the human species. Knowledge of the future is of value to everyone. In the past when a demand for knowledge of the future existed, mystical forms of prophecy came into existence. Priests, witches, prophets, crystal balls, astrology, palmistry, and oracles all played a part.

## THE GREAT PYRAMID OF CHEOPS

There exist sites and remains of structures that have played important roles in predicting the future in ages past. One of the earliest is the Great Pyramid of Cheops (ca. 2650 B.C.). The size and finesse of construction of this pyramid more than 4000 years ago has led to speculation of every kind about its construction and what it means. It is a monument to Pharoah Cheops, founder of the fourth dynasty. The base of the pyramid covers 51,000 square meters and the apex stands 147 meters above the ground. A mass of laborers constructed the pyramid. Perhaps as many as 100,000 laborers built the monument. They moved more than 2 million stone blocks down the Nile River to near Cairo. They then transported the blocks to the west side of the Nile valley and hoisted them onto the escarpment. There they assembled the blocks into the structure that remains today. The average weight of the blocks is 2.5 metric tons. They then fitted white limestone pieces so as to provide a smooth surface to the structure. Most of the white facing is now gone. It was probably pirated over time for other structures. Only a few pieces still remain near the top. They are visible in Figure 17.1.

Inside the structure are a series of passageways that lead to two burial chambers, one for the pharaoh and the other for his wife. Some pyramidologists indicate that in the construction of the passageways is a basic history of the world from 4000 B.C. to the year A.D. 2001. A Scottish astronomer, Charles Piayyi Smyth, made accurate measurements of the direction and dimensions of the passageways in 1864. Based on his measurements he came up with a chronology covering 6000 years. He used one pyramid inch (25.25 millimeters) to represent one year. Downturns and restrictions in passageways represent hard times and world disasters. Upturns, broad passageways, and the burial chambers themselves represent good times and major advances for the human species. Some of the structural chronology and significant world events coincide. However, either the human species did not heed the message, or there were mistakes made in construction because the system fails frequently. They built the passageways as they are for real reasons. Certainly, a people capable of the design and construction of the monument did not build the interior randomly. However, their reasons are now unknown. The end of the corridors implies a great new world by 2001, an optimistic prediction that we hope will be correct.

**Figure 17.1** The Great Pyramid of Cheops. The pyramid stands on the desert plateau just outside of Cairo, Egypt. It was completed about 2650 B.C.

### *THE ORACLE OF DELPHI*

High on the side of Mount Parnassus in Greece is the site of Delphi. Early Greeks built temples to Apollo and Dionysus there, and a famous and powerful oracle came into existence there. An oracle was a medium through which a pagan god revealed divine knowledge about the future, or outcomes of events. The oracle existed for perhaps 300 years beginning in about 700 B.C. The early temple was a crude one of wood and feathers, but with time, they built larger and more elaborate stone temples (Figure 17.2). The Greek people of the time believed that Delphi was at the center of the earth. In the beginning, prophecies were made on only one day a year. As the fame and power of the oracle grew, it became necessary to make prophecies one day a month. So influential did the oracle become that cities began to build their own temples on the site. Part of the reason for the success of the oracle was that the prophecies were always correct. The oracle accomplished this by making prophecies which were so vague that the priests who interpreted the prophecies could always show why they were right. Due to the high rate of success, kings and

**Figure 17.2** The Archeological site of Delphi. The site is high on the side of Mount Parnassus in central Greece.

city-states paid large sums for prophecies. Officials of Greek city-states frequently consulted the oracle before they began a major venture such as a new colony or a war.

King Croesus of Lydia questioned the oracle about his proposed war on the Persian Empire in 550 B.C. A huge sum was paid for a prophecy, which King Croesus believed said that he would win the war. The fee included a gold lion weighing 259 kilograms, a gold statue of a servant nearly 1.5 meters in height, and 117 bricks of assorted precious metals. The price was higher than expected, as it cost him his empire as well. The oracle later came under ill-use from the Romans, and the power of Delphi declined.

Exactly how and why Delphi came into existence is not clear. Whoever selected the site for Delphi provided half the success. The site itself is awe inspiring. It is situated on the slopes of the mountain with a deep green valley below. The blue waters of the Mediterranean Sea are visible in the distance, as are the often snow-capped peaks in the distance. It would inspire anyone with any aesthetic appreciation. The site coupled with the infallibility of the messages delivered ensured its fame.

## *SCIENCE AND FORECASTING*

Some oscillations repeat themselves so many times and so regularly that the rhythm does not appear to be the result of chance. This information is used to forecast behavior in the future. The regularity in some natural events is much more easily discerned than others. Periodic events associated with the movements of the planets are among those forecast with considerable accuracy. Success in forecasting periodic events developed first in astronomy. The regularity of the movements of the moon and planets led to prediction of astronomic events well in advance of their occurrence. In many ancient societies, the ability to forecast seasonal floods, tides, eclipses, and comets gave the priesthood or ruling class major power. The discovery in Mesopotamia of the association of the annual floods with a certain arrangement of the stars allowed the priests to forecast the annual floods of the Tigris and Euphrates rivers. The mystical concepts of astrology probably derive from this period in history.

The present movements of the moon and planets will continue barring a major cosmic disaster. There will be the daily ebb and flow of energy around the earth as the earth spins on its axis. Along with this ebb and flow will be the daily changes in atmospheric chemistry that take place as photosynthesis and respiration alternate in the world of green plants. Land and sea breezes and mountain and valley breezes will continue to follow this daily flow of energy. The tides will continue their progression through time as the earth and moon move through their orbits. The seasons will be the dominate force for life on the planet, never being quite the same from one year to the next, but changing only slowly through time as the planet changes its position in space. These regular oscillations will continue to be the regulators of the biosphere.

## *THE NEED FOR FORECASTING*

There are many questions about the future global system for which we need information. One entire group centers around the widespread and varied impact that climate change would have on other aspects of the environment. Among the many things that would change if climate changes are sea level, biological productivity, and biological diversity both on land and in the ocean. Some notion of the difficulty of forecasting global environmental change is the complexity of the interaction and feedback between various parts of the global system. For example, human-induced increases in $CO_2$ and other trace gases are major elements in potential global warming. However, because $CO_2$ is the primary raw material for photosynthesis, increased $CO_2$ concentration is likely to have a direct biological impact on the extent and distribution of Earth's vegetal cover. However, changes in the vegetal cover changes the hydrology and reflectivity of the surface. This might further affect climate.

Human-induced changes in climate and nutrient balances will influence biological productivity and biodiversity. Acid deposition, surface ozone, and oxident levels will influence biological production and diversity. In the seas, low dissolved oxygen and addition of chemicals will affect biodiversity and production. Land-use changes such as deforestation and irrigation will affect both atmospheric concentrations of trace gases in the atmosphere and nutrient balances between terrestrial and marine ecosystems.

## *RECENT DEVELOPMENTS IN FORECASTING*

The desirability—in fact, the necessity—of planning in modern life has led to the development of techniques that can be used to provide some intelligent basis of forecasting future events upon which planning can be based. Among these are various types of models and scenario construction.

It is not possible to forecast future environmental change by conducting direct experiments. To forecast global change, scientists must use other means. One such method is the use of models. A model is a simple representation of something real. A model car, for example, has the general shape and characteristics of a car but may not have a motor or any working parts. A toy model car does not need to be like the real thing except in basic shape. It is also true, however, that physical models such as the toy car serve as analogs to obtain important predictive information. A model car might be used in a wind tunnel to determine the way in which air flows around a real car. Not all models are physical models like a model car. There are also graphic models, conceptual models, statistical models, and mathematical models.

### Analog Models

Analog models have attributes that are in some ways similar to conditions that exist in the real world. A classic example is the electrical analog model, which uses voltages, variable resistors, and amplifiers to represent various parts of a system. In an analog model of climate, a constant flow of electricity represents the flow of energy from the sun. Resistors modify the current and represent atmospheric components such as clouds or water vapor.

It is perhaps natural that researchers should look for past global climates to see if there are any guides to the future. An integral part of global environmental research is finding previous global environments that have particular sets of conditions. Of particular importance is the identification of earlier times when a warmer climate existed. The environmental conditions that prevailed at those times could prove helpful in estimating potential future conditions if global warming exists.

Analysis of paleoclimates warmer than our own is now taking place. There are three periods in the past that may be analogs for Earth subjected to global warming.

1. The climatic optimum that occurred about 5000 years ago. This warm time saw summer temperatures in high latitudes some 3 to 4°C above current levels. There was increased precipitation in subtropical and high latitudes. Data from the Cooperative Holocene Mapping Project (COHMAP) suggest that during this climatic optimum in middle latitudes, summer temperatures were only 1 to 2°C higher than now. These areas

also had increased annual precipitation. Although there are uncertainties in this reconstruction, it does provide a relative guide to warmer conditions.

2. The Eemian Pleistocene interglacial existed from 125,000 to 130,000 years ago. This interglacial experienced very warm conditions compared to other interglacials.
3. The Pliocene warm climate that occurred about 3.3 to 4.3 million years ago. This warm period was one of the wide swings between warm and cool climates that occurred during the Pliocene.

### Simulation Models

Simulation techniques is a term for a variety of analytical techniques. Among them are systems analysis, mathematical modeling, operations research, linear programming, and dynamic programming. Simulation is the development and application of mathematical models to represent the interaction of physical processes through time. They can be used to produce future states of a system. Simulation techniques first appeared around 1959. Since then the practical use of these mathematical models increased rapidly. The rate of growth in the use of simulation techniques has paralleled the advance in computer technology.

In principal, a model is a generalized image of a physical system. A model imitates the real system so that new information can be obtained about the system. Models are different from the real world, hence they are analogs. They are selective approximations which by the elimination of incidental detail allow some basic, relevent, or interesting aspects of the world to appear in generalized form. Because they are selective in character, they all have limitations.

### Mathematical Models

Mathematical models consist of equations that simulate environmental processes. When the model is run with appropriate data, it yields the state of the system at a future time. All such models are simplifications of the real world. There are several reasons for this. First, scientists do not understand well enough the processes that bring about environmental change. There is much yet to learn about the size and rate of changes. A second factor is that even the fastest computers in operation today cannot deal with such a complex system over such a large space. Computers with the speed and capacity required to simulate Earth processes as a single system do not exist. Therefore, simplified representations of many important processes must be used. The end result is that only pieces of the global system can be predicted for the future.

### Climate Models

Some global models simulate important processes of global change. Among these are the global circulation models (GCMs) used to forecast future climate. Probably the most useful aspect of these models is that they provide geographical information about the variables under study. Presently, they do not provide very detailed geographic analysis. Typically, a point in the model may represent an area the size of the state of Colorado. The GCMs can predict changes in temperature, precipitation, wind, snow accumulation, and soil moisture over units several hundred kilometers square.

Modern GCMs depend on high-speed computers with prodigious memory. Even attempting to recreate the complexities of atmospheric circulation would be impossible without these machines. A GCM attempts to solve (1) equations of motion, (2) thermodynamic equations, and

(3) conservation equations for moisture and mass. The models operate for the following geographical space: (1) a defined global area, and (2) through various atmospheric levels.

The model used at the National Center for Atmospheric Research (NCAR) has a grid with nine layers stacked to a height of about 30 kilometers. Spacing between the centers of the rectangular grid points is about 4.5° of latitude and 7° of longitude. Even with this coarse network it takes several days for a computer run. A moderately complex GCM predicting future climate in yearly steps for 100 years may take two months. If the size of the grid were reduced to one-tenth of the existing size, it would take a supercomputer many years to derive the result.

An average value represents the entire rectangular grid point. If there are significant departures from the average within a given grid unit, the model cannot represent changes for those areas. The size of each grid point is greater than some mesoscale weather systems, so it ignores these major storms. Also, predicting changes in cloud types and amount of cloud cover is not yet possible.

## GLOBAL ECONOMIC MODELS

In 1972 came the publication of the results of the first simulation model of the global economic system. A futures group known as the Club of Rome produced the study. The title was *The Limits to Growth* (Meadows et al., 1972). There was global response to the report. By some it was taken as a projection of doom for the earth as we know it by the year 2100. The report showed what the future would hold if we did not interfere with present trends. It concluded that there are limits to the size of human population which can be sustained on Earth at any given economic level. Changes in our value system can bring about the changes necessary to prevent the disaster forecast in the model.

As a result of criticism of the first report, the Volkswagon Foundation funded a second study. This report was based on a much improved computer model. It also divided the world into different regions rather than treating it as a single region. The results differed some and there were some modifying clauses. However, the final conclusions were similar to the earlier study. Essentially, both said that if changes are not made, the human species would have increasingly severe environmental problems.

Soon after, Kahn and others (1976) provided a completely different picture of the future. He predicted some near-term problems and uncertainties for the species but a beautiful long-term picture. He rejected the premise of physical limits to growth in his scenarios of the future. The studies by the Club of Rome and by Herman Kahn project the future from 2020 to 2150. These reports agree in some areas. Two areas of agreement are that there is a limit to population growth and a limit to economic growth. There is no agreement whether there are physical limits to growth or when we might reach such limits.

Despite the many attacks on *The Limits to Growth*, the thesis of the book persists. No one can dispel the basic forecast presented. There is too much evidence that the prediction may eventually come true. Coupled with this is a growing, perhaps fully warranted fear that humans are setting in motion forces of a scale of space and time that we can no longer control.

### United Nations Conference on the Human Environment

The United Nations recognized the growing problem of human relationships with the environment and convened a conference in Stockholm in July 1972. One of the accomplishments of the conference was a declaration of principles and guidelines (Appendix A). The declaration was significant because (1) it recognized the Earth as a single physical and biological system, and (2) there is a worldwide problem in the human interaction with the environment.

Frequently overlooked in the discussions of changing consumer preferences, employment opportunities, and pollution loads is that these concepts are irrelevant to most of the human population. These concepts have meaning only in the rich countries where such choices exist. The projections for a century or more in advance have no meaning where life depends on a single season's rainfall. It is painfully clear also that the gap between the rich and the poor countries is increasing.

### The Global 2000 Report to the President

In May 1977, President Carter issued a directive for a study of global population, resources, and environmental change to the year 2000. The study was to become the base for long-range planning. The three-volume report published in 1980 was *The Global 2000 Report to the President: Entering the 21st Century.* It represents the first try by the U.S. government to produce a consistent set of projections for the Earth, integrating population, resources, and environmental impact. The study projected a variety of phenomena to the year 2000. The study group made the assumption that present trends and policies would continue. They could find no evidence that long-term trends would change. Experience with the legislative process suggested that policies probably would not change either. The results of this study, when summarized, are basically the same as those of *The Limits to Growth.*

Perhaps Loye (1978, p. 24) sums it up when he states: "The futurist projects the logical multivariate patterns that growth can take, and poses their alternative consequences, in order to place before the decision makers the choices that must be made if desired, rather than undesired, futures are to be realized."

## *SCENARIO DEVELOPMENT*

Another method of trying to forecast the future is the scenario. It combines logical reasoning with speculation and imagination to produce a future situation. The scenario identifies a situation, assumes some event, and then examines the effects of that event on the system or situation. A thoroughly done scenario takes the entire system into consideration. It includes the side effects as well as the primary effects of the event.

Perhaps one of the earliest uses of the scenario was by G. G. Chesney in 1871. He published a work called "The Battle of Dorking" in *Blackwood's Magazine.* In the work he forecast a German invasion of Britain and a following battle that Britain lost due to lack of preparedness. Scenarios are possible sequences of events constructed to bring attention to the results of certain processes at work. To be effective the sequence of events must be plausible and realistic. Today, if the results are to be at all credible, the scenario must cover as many likely developments as possible. Scenarios are being used today to assess the impact of future global environmental changes. They provide possible futures that may serve as guides to planners, with some limitations that go with all forecasting techniques.

## *SUMMARY*

As the human population grows and the world enters further into a global economy, forecasting future events becomes ever more important. There is no way of knowing the future for certain. There are a variety of forecasting techniques that can be used to provide possible and probable events. These include the use of models of various types, of which computer simulation models are among the more sophisticted. Simulation models are now being used to forecast natural changes in the physical environment as well as economic and social events. The more accurate the models are, the more reliance can be placed on them for planning purposes. Planning for the future is a necessity for the human species.

## BIBLIOGRAPHY

KAHN, H., AND OTHERS 1976. *The Next 200 Years: A Scenario for America and the World.* New York: Morrow.

LOYE, D. 1978. *The Knowable Future: A Psychology of Forecasting and Prophecy.* New York: Wiley.

MEADOWS, D. H., D. L. MEADOWS, J. RANDERS, and W. W. BEHRENS. 1972. *The Limits to Growth.* Washington, D.C.: Potomac Associates, University Books.

U.S. COUNCIL ON ENVIRONMENTAL QUALITY AND THE DEPARTMENT OF STATE. 1980. *The Global 2000 Report to the President of the U.S.: Entering the 21st Century.* Vol. 1. The Summary Report. Washington, D.C.: U.S. Government Printing Office.

# chapter 18

# The Potential for Global Warming

***CHAPTER SUMMARY***



Life is possible on planet Earth solely because of a process known as the greenhouse effect (Figure 18.1). If the planet had no complex interchange of energy among the ocean, atmosphere, lithosphere, and biosphere, temperatures over the planet would be too cold to support life. Temperature observations begun in the mid-nineteenth century show that globally averaged surface temperatures increased 0.3 to 0.6°C over the past century. Most of the increase took place in the 1920s and 1980s. The 1980s were the warmest decade on record. Nineteen eighty-eight was the warmest single year on record. Both earth- and satellite measurements support the most recent global changes in temperature. Some scientists suggest that the cause for the warming is in an increase in atmospheric carbon dioxide and other gases resulting from human activities.

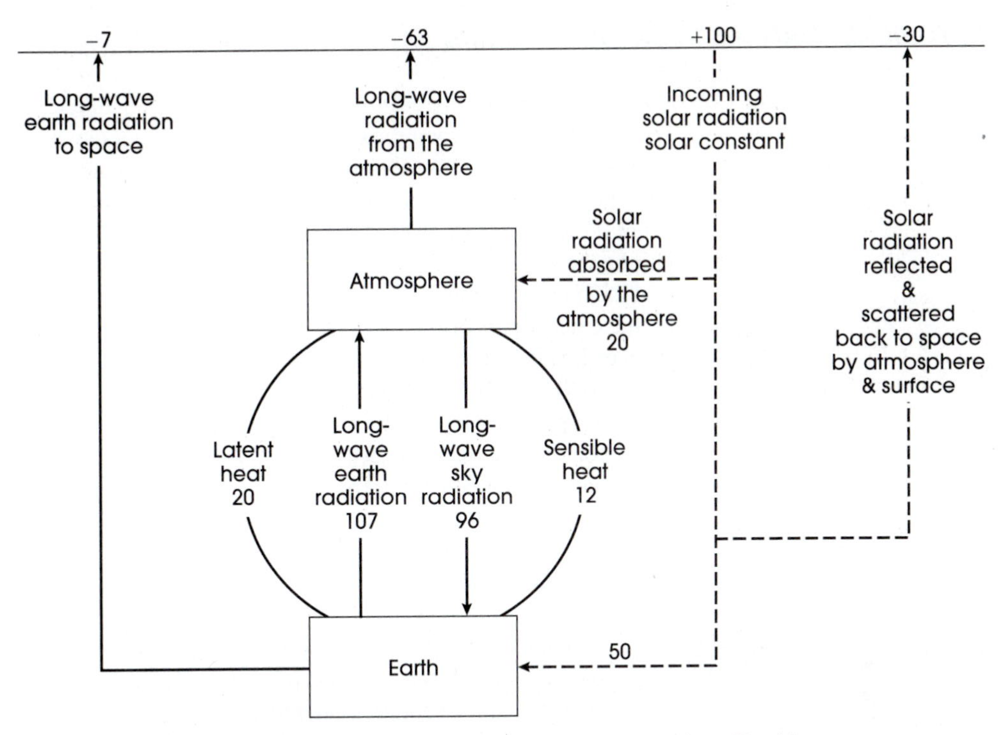


**Figure 18.1** The greenhouse effect of the atmosphere raises Earth's temperature from below 0°C to about 15°C. The increased surface temperature is due to the cycling of energy back and forth between Earth's surface and the troposphere.

Carbon dioxide ($CO_2$) is one of the many variable gases found in the atmosphere. Atmospheric $CO_2$ is an active trace gas. Compared to the major constant gases of nitrogen and oxygen, it is found only in small quantities. It averages less than $\frac{1}{2}$ of 1 percent of the atmosphere by volume (about 350 ppm). The amount of carbon dioxide in the atmosphere varies with time. There are regular oscillations in the $CO_2$ content on a daily and seasonal basis. During the day, photosynthesis withdraws $CO_2$ from the atmosphere, and at night respiration releases $CO_2$. Daily fluctuations during the growing season in midlatitudes are as high as 70 parts per million (ppm).

There is also a seasonal change in $CO_2$ in midlatitudes resulting from variation in the rate of photosynthesis. The carbon dioxide content rises to a peak in spring in the northern hemisphere and falls to a minimum in late September or October. The seasonal change in atmospheric carbon dioxide reflects a very important factor affecting the atmosphere: the metabolism of all living matter. The seasonal change in carbon dioxide concentration in the atmosphere results from the "pulse" of photosynthesis. This occurs during the summer in midlatitudes of both hemispheres. The seasonal difference is only about 5 ppm in warm humid areas such as Mauna Loa, Hawaii. It is more than 15 ppm in central Long Island, New York. The difference is less near the equator, where the vegetation is green most of the year. It is largest in midlatitudes, where the vegetation is largely deciduous. The difference is also less at higher elevations at all latitudes. These diurnal and seasonal changes in $CO_2$ have been a part of the natural atmosphere over geological time. There are also irregular changes that take place over longer periods. These are due to changes in volcanic activity, changes in the rate of chemical weathering, and in volume of the living vegetation on Earth.

## HISTORICAL CHANGES IN CARBON DIOXIDE

The $CO_2$ content of the atmosphere dropped steadily through geologic time until about 50 million years ago. The levels of carbon dioxide in the atmosphere today have existed for the past 3 to 4 million years ago. Ice cores taken from Antarctica and Greenland now go back 160,000 years, with the longest taken at Vostok in the antarctic. Contained in the ice of these cores are small bubbles of air from the atmosphere at the time the bubbles formed. Analysis of these air bubbles shows that during the last 2 million years carbon dioxide varied over a relatively small range, between 180 and 280 ppm (Chapter 4).

During the coldest part of the last glaciation, carbon dioxide dropped to about 180 ppm, with a variation of less than 10 ppm. This level of carbon dioxide is about 30 percent lower than during the past 10,000 years. In the 1000 years preceding 1850, the concentration of carbon dioxide stayed near 280 ppm by volume. The most likely reason for this fluctuation is that cold water absorbs more carbon dioxide than does warm water. As a result of this process, during periods of glaciation the carbon dioxide content drops.

## INCREASE IN CARBON DIOXIDE SINCE 1850

Data show the $CO_2$ content of the atmosphere has been increasing since at least 1850 (Figure 18.2). Just how much the total increase has been is uncertain, due to the nature of the early measurements. Early in the twentieth century scattered measurements showed an average annual concentration of 316 ppm. Continuous data are available only from 1958, when Charles D. Keeling

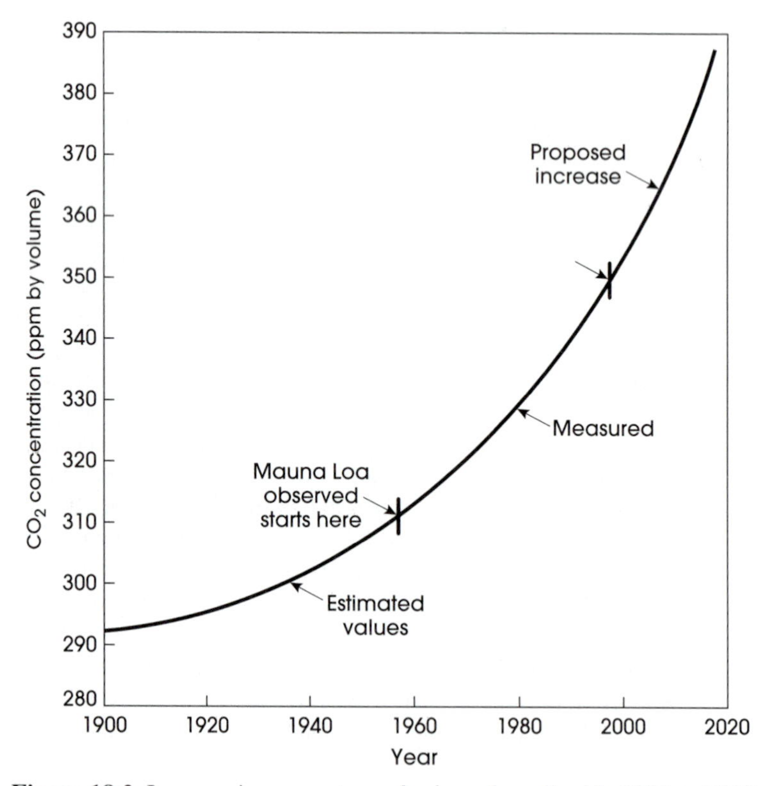


**Figure 18.2** Increase in mean atmospheric carbon dioxide 1900 to 2020.

of the Scripps Institute of Oceanography set up a continuous monitoring station at Mauna Loa on the island of Hawaii. The $CO_2$ concentration increased from about 280 ppm in 1850 to about 353 ppm in 1990 (Figure 18.3). The amount of increase at Mauna Loa averaged 0.8 ppm per year. The increase ranged from 0.5 to 1.5 ppm per year.

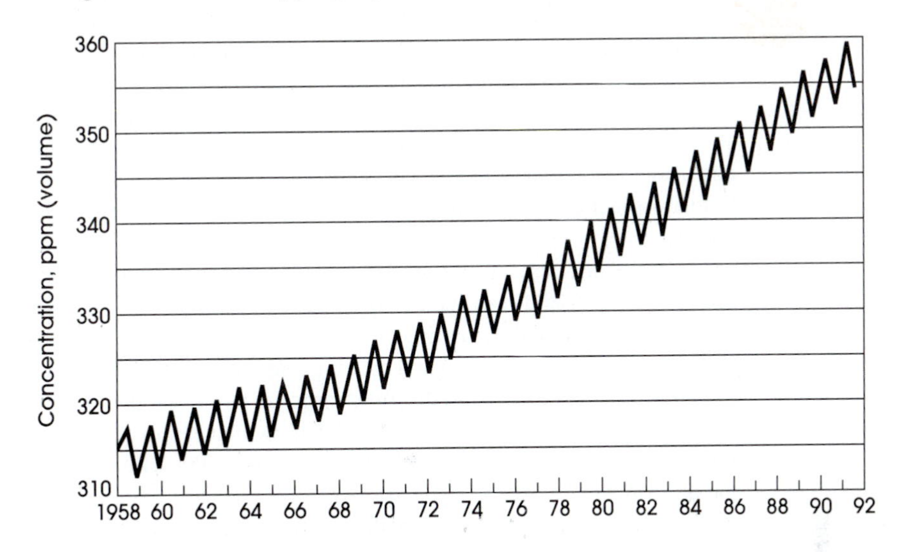

**Figure 18.3** (a) Changes in atmospheric $CO_2$ between 1958 and 1992 at Mauna Loa Observatory, Hawaii. (b) Sea level rise due to recent warming and local tectonic activity have drowned ancient Roman and Minoan fish tanks between the islands of Crete and Mochias.

A significant aspect of this change is that the rate of change is increasing. From 1958 to 1968 the increase averaged 0.7 ppm per year. From 1968 to 1978 the average increase was 1.3 ppm per year. A fourth of the total increase occurred in the decade 1967–1977. In 1977–1978 the increase was 1.5 ppm for the whole atmosphere and 2.0 ppm over the antarctic continent. The $CO_2$ concentration of 353 ppm in 1990 is greater than at any other time in the past 100,000 years ago. The concentration is now 26 percent greater than in the preindustrial era from 1750 to 1800. It is increasing about 1.8 ppm (0.5 percent) annually because of human activities.

The main sources for the increased $CO_2$ are the combination of burning fossil fuels and destruction of natural vegetation to clear land for agriculture. Over the past century, fossil fuel use and cement manufacturing released about 200 billion tons of carbon into the atmosphere. In 1987, human-influenced emissions of carbon dioxide from fossil fuel use were about 5.7 gigatons of carbon.

In the last century, deforestation may have released as much as an additional 115 billion tons of carbon. Land-use changes in 1987 added another 1.6 to 2 GTC to the atmosphere. Land-use practices reduce the standing crop of living biological matter through deforestation and burning vegetation. Deforestation and burning vegetation cause the direct release of carbon dioxide to the atmosphere. They also reduce the removal of carbon dioxide from the atmosphere. Removing live vegetation reduces photosynthesis, which removes carbon dioxide. Deforestation and burning vegetation may also disturb soil processes that affect the carbon dioxide exchange with the atmosphere.

There are natural feedbacks that work to control carbon dioxide content in the atmosphere. For example, the ocean is a major absorber of atmospheric carbon dioxide. Colder water absorbs more carbon dioxide than does warm water. Thus the amount of change in temperature depends on the rate at which natural feedback mechanisms react to lower the carbon dioxide content. The faster the feedback mechanisms operate, the less the temperature change is likely to be. Some 40 percent of the gas placed in the atmosphere was absorbed either by Earth's biomass or by the ocean. The remainder is in the atmosphere.

## *THE CARBON CYCLE*

Of the estimated 315 billion tons of carbon placed in the atmosphere since 1850, only 130 billion tons remain there. During the last decade (1980–1990), about half of the human contributed emissions stayed in the atmosphere. One of the biggest questions currently is: Where has the remaining carbon gone, and what processes removed it? There are at least two places where atmospheric carbon goes: into the oceans and into green vegetation. The oceans contain about 50 times more carbon than does the atmosphere. The ocean is an important long-term sink for carbon from the atmosphere. The amount of carbon the ocean takes up and releases depends on a variety of physical, chemical, and biological processes. We understand only some of these.

There is an exchange of carbon dioxide between the atmosphere and vegetation on land that is about the same amount as that between the atmosphere and the seas. This is about 100 gigatons per year. Terrestrial vegetation removes atmospheric $CO_2$ through photosynthesis. Vegetation and soils in northern midlatitudes may be becoming more effective in removing carbon from the atmosphere because: (1) there is an increasing amount of land in forest, (2) $CO_2$ stimulates vegetative growth, and (3) vegetation uses more $CO_2$ in warmer weather.

Uncertainties in the size of individual sources and sinks of $CO_2$ severely limits the accuracy of forecasts of future atmospheric concentrations. The natural movement of $CO_2$ into and out of the oceans and terrestrial vegetation is much larger than from human activity.

The task of predicting future abundance of atmospheric $CO_2$ requires scientific information from many scientific disciplines. We must understand how the $CO_2$ budget operates today. We also need to know how it responds to changes in climate and other environmental conditions.

Although forecasting is not reliable, present data suggest that $CO_2$ emissions will grow by about 1.6 percent annually from 1980 to 2025 and then decline to a growth rate of around 1 percent per year. Based upon present rates of change it is probable that atmospheric $CO_2$ will rise 32 percent above the 1850 level by A.D. 2000. It is both possible and probable that at some time the level will reach 400 ppm. Once the concentration reaches 400 ppm, it will be at a level not attained in the past million years. This level is significant even compared to geological time. The volume of coal known to exist in reserves is far more than necessary to produce such an increase in $CO_2$. The amount of atmospheric $CO_2$ may well double from the 1980 level in the years between 2050 and 2100. If all the known reserves of fossil fuels are burned, the $CO_2$ content in the atmosphere would triple. The earth is now as warm as it has been any time in the past 125,000 years. If the $CO_2$ content increases, so will global temperatures. The predictions of how much temperatures will increase vary significantly. The estimates vary because the mathematical models of atmospheric processes are not complete. If $CO_2$ concentrations double from the preindustrial level of 280 ppm, mean world temperatures will increase anywhere from 1.5 to 4.5°C. The National Research Council (1983) placed the temperature range from a doubling from 300 ppm to 600 ppm at 2 to 3.5°C. If mean global temperatures increase, the warming will not be the same every place on the globe. The subpolar latitudes will experience a much greater warming than equatorial regions. These subpolar regions will warm two or three times as much as equatorial regions. In the northern hemisphere is a zone where the snow cover shifts north and south from year to year. In this zone, mean temperatures will increase 4°C if $CO_2$ content doubles. There is no evidence now of any changes in the cover of Arctic sea ice or snow cover on the adjacent landmasses.

Actual temperature changes do not bear out the impact of the increasing $CO_2$ content. The amount of fuel burned to date suggests that the earth should have undergone a 0.5°C rise in temperature. This has not taken place. Part of the problem is that there are other factors which cause the temperature to oscillate independently of the carbon dioxide in the atmosphere. The lack of warming in recent years has led some people to give less significance to the warming effects of the $CO_2$. If other variables do not play a part, the exponential rise in the atmospheric $CO_2$ content will become significant. Future drops in temperatures will only moderate the effect of the $CO_2$.

Based on the projected increase in $CO_2$, the mean temperature at the surface of the earth may increase 2.8°C, which would result in heating the surface of the oceans by the same amount. It will take at least 50 years to warm the top layer (100 meters) of the ocean and a minimum of 250 years to adjust the temperature of the bottom waters. Assuming the present deep-sea circulation, it would take 1000 years or more to bring the ocean into adjustment with a 2.8°C rise in surface temperatures.

## *CARBON DIOXIDE AND THE THEORY OF GLOBAL WARMING*

An increase in carbon dioxide has considerable potential for changing the earth's energy balance. Carbon dioxide is partially responsible for the "greenhouse effect" of the atmosphere, as it is transparent to solar radiation but quite opaque to infrared radiation. Earth's radiation is concentrated in the infrared band from 7 to 20 micrometers. Carbon dioxide is absorbent of radiant energy in the wavelengths in which the earth radiates heat away from the surface, and it is in the range 15 to 20 micrometers, where most of the absorption takes place (Figure 18.4). If carbon dioxide absorbs earth radiation and the amount of carbon dioxide in the atmosphere is increasing,

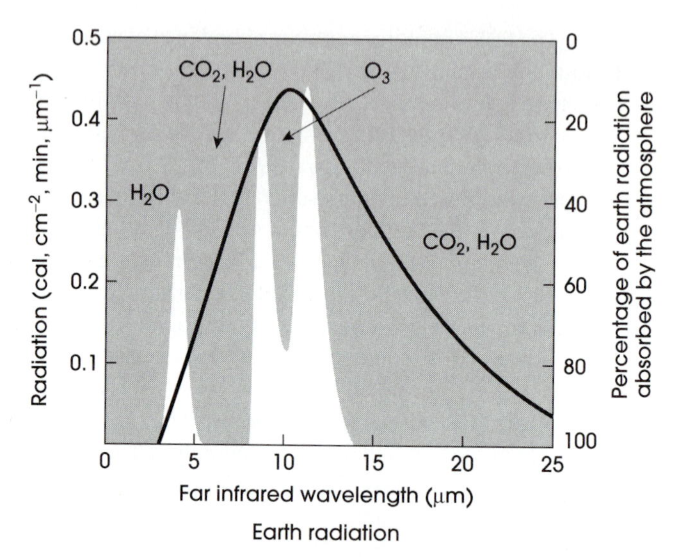


**Figure 18.4** Atmospheric $CO_2$ is a major absorber of earth radiation. It absorbs radiation over most of the range of wavelengths at which earth radiation occurs.

more earth energy should be absorbed by the atmosphere. This absorption and reradiation back to the surface will shift the energy balance toward increased storage of energy, hence raising the temperature of the earth's surface and atmosphere.

Natural changes in temperature due to unknown interannual variations are as large as changes predicted due to changes in $CO_2$. It is a complex process. All the increase in temperature in the past 100 years may be due to natural processes. During the last interglacial, it was warmer than it is now. It is also possible that global warming due to $CO_2$ forcing is much larger than indicated by mean Earth temperatures. Other processes may be cooling the atmosphere. Whether or not global warming due to $CO_2$ forcing exists is still an open question.

Estimates of how much the temperature will increase vary significantly. The estimates vary because they come from different mathematical models of atmospheric processes. If $CO_2$ concentrations double from the preindustrial level of 280 ppm, mean world temperatures will probably increase about 1.5 to 4.5°C (3 to 8°F). The National Research Council (1983) placed the range for a doubling from 300 ppm to 600 ppm at 2 to 3.5°C (4 to 6°F). If mean global temperatures increase, the warming will not be equal over the globe. The subpolar latitudes will probably experience a much greater warming than will equatorial regions. These subpolar regions will warm two or three times as much as equatorial regions. In the northern hemisphere, the zone where the seasonal snow cover shifts from year to year, mean temperatures may increase 4°C (7°F) if $CO_2$ content doubles.

## *CONTROVERSIAL EVIDENCE FOR GLOBAL WARMING*

In testimony to the U.S. Senate on June 23, 1988, James Hansen asserted that it could be said with about 99 percent confidence that Earth is getting warmer. Others began to support the hypothesis of global warming. The support was based mainly on three pieces of evidence: (1) 1990 was the warmest year of record, (2) the four warmest years of record were in the 1980s, and (3) the decade of the 1980s was the warmest decade of record.

On the basis of historical data for the period for which records have been kept, there is no cause for alarm (Figure 18.5). Actual temperature changes do not support the impact of the increasing $CO_2$ content. The earth should have undergone a 0.5°C (0.9°F) rise in temperature based on the amount of fuel burned to date, but this has not taken place. Careful studies of the historical record show that the change in temperature that has occurred since the beginning of the industrial revolution does not vary significantly from zero. The year 1880 marks the beginning of the historical record for which there is enough data to provide credible information. Critical study of the data show that probably not even 0.5°C warming has actually taken place (Figure 18.6). For example, urbanization has increased the temperature at about a third of the weather stations with long records. In addition, most of the earth is covered with an ocean for which there are few

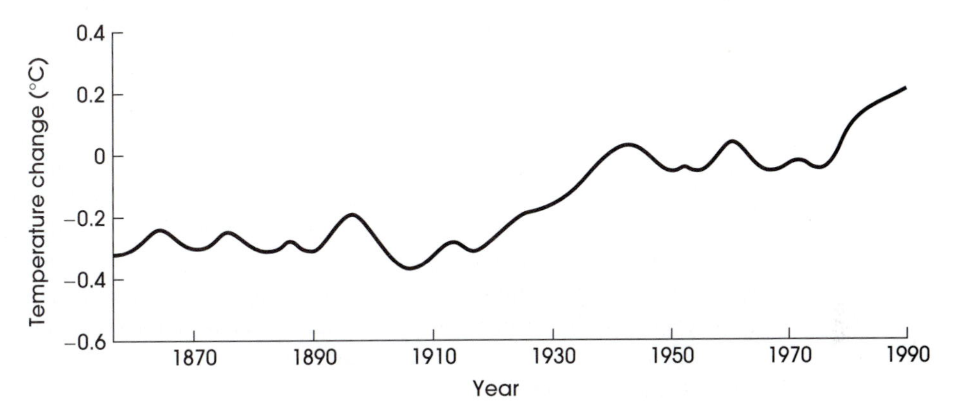


**Figure 18.5** Globally averaged temperatures since about 1860 shown as a change from the 1951–1980 average.

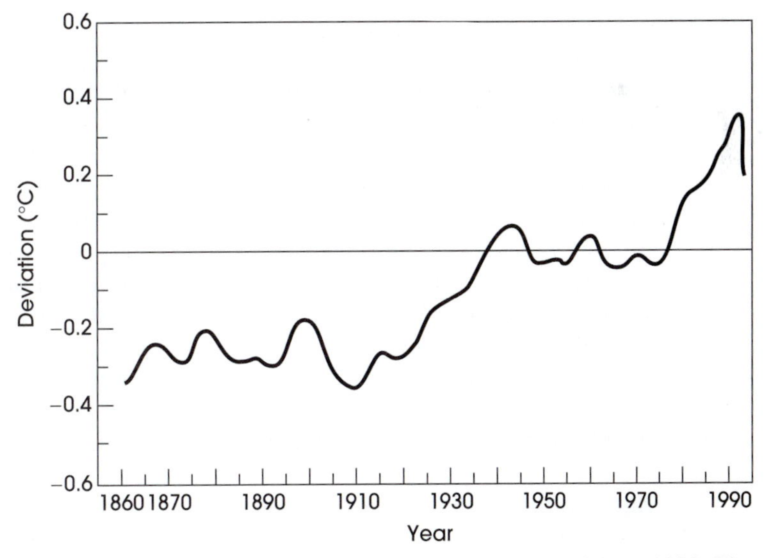


**Figure 18.6** Combined land, sea, and air temperatures from 1861 to 1992. The temperatures are expressed as departures from the mean of the climatic normals for 1951–1980.

surface temperature data. Global circulation models predict the most warming for the region around the Arctic Basin. Such warming has not yet been detected.

## *COMPLEXITY OF THE ENERGY BALANCE*

The problem of global warming due to carbon dioxide is a complex one. Many factors affect atmospheric carbon dioxide. There are natural feedbacks that work to control $CO_2$ content in the atmosphere. One of these feedback mechanisms is that carbon dioxide is soluble in seawater. Thus the change in temperature that might occur depends to some extent on the rate at which natural feedback mechanisms react to lower the $CO_2$ content. The faster the feedback mechanisms operate, the less the temperature increase is likely to be. The earth's biomass and the ocean absorbed some 40 percent of the gas placed in the atmosphere. The remainder has accumulated in the atmosphere.

Part of the problem is that other factors cause the temperature to oscillate independently of the carbon dioxide in the atmosphere. One atmospheric component that affects global temperature is pollutants in the form of particulates. One of these pollutants is sulfur dioxide. Sulfur dioxide particles serve as very small nuclei for cloud particles. Clouds made of very fine particles reflect more solar radiation than do clouds made of large particles. Since sulfur dioxide and carbon dioxide are both products of burning fossil fuels, they work in opposite directions on the energy balance. One leads to warming and the other to cooling. It is possible that increased reflectivity of the atmosphere may be offsetting as much as half of the potential greenhouse warming.

Other variables being constant, $CO_2$ content in the atmosphere will continue to increase for the foreseeable future. On the other hand, continued increases in $CO_2$ will have less effect on global temperatures than previous ones. Each additional increment of $CO_2$ has less effect on the energy balance than the preceding one.

## *INCREASES IN METHANE*

Other gases also play a part in global warming. Methane is a gas found in small amounts in the atmosphere under natural conditions. Methane is a greenhouse gas. It is very absorbent of Earth radiation. Evidence shows that it has begun to accumulate during the past two centuries. Swiss scientists used gases trapped in ice cores from Greenland and Antarctica to study changes in methane. They found that 100,000 years ago methane was present in the atmosphere at about 500 parts per billion by volume (ppbv). Seventy thousand years ago it was about 650 ppbv. By 20,000 years ago, at the height of the last glacial advance, the level dropped to 350 ppbv. With Holocene warming the level climbed to near 650 ppbv once again. The concentration stayed near this level for the 3000 years prior to 1800. In the past 200 years levels have risen 250 percent. In the latter half of the nineteenth century, methane was present at about 800 ppb. This increased to about 1600 ppb in the mid-1990s. About half of the increase came since 1960. The additional methane comes from livestock raising, rice cultivation, industry, mining, and landfills.

### Methane and Noctilucent Clouds

Methane rises in the atmosphere passing through the troposphere to the stratosphere. In the stratosphere methane breaks down under sunlight and the hydrogen oxidizes to form water molecules. Water vapor is normally not present in any significant amounts above the troposphere. In the stratosphere the amount ranges up to about 3 ppm. Each methane molecule produces two water molecules. Computer models show that a doubling of methane in the atmosphere should increase water vapor in the stratosphere by some 30 percent. This is enough water vapor to produce clouds made up of water and ice crystals at the cold temperatures found at 85 kilometers. This places them high in the stratosphere. The water which condenses to form these ice crystals comes from the breakdown of methane.

On the night of June 8, 1885, T. W. Backhouse recorded the first observation of a noctilucent cloud. Soon many more sightings were reported. These clouds usually appear around midnight and are most frequent in the three weeks before and after the summer solstice. They are particularly visible between latitudes 50° and 60°. The noctilucent clouds are created by human activity through increasing methane in the atmosphere.

### Methane and Permafrost

In latitudes from about 50 to 60° in the northern hemisphere much of the land surface is covered by permafrost. Permafrost is ground that is frozen in winter to considerable depth. In summer the surface melts, but it does not melt down far enough to thaw all the soil. The permafrost provides a barrier for gases beneath and water above. There is a lot of methane trapped beneath this frozen soil. Because global warming may be quite large in subarctic regions, rapid melting of the permafrost may take place and this will release methane into the atmosphere. Methane is about 20 times more effective than $CO_2$ in trapping Earth radiation. If more methane is released into the atmosphere there may be a positive feedback effect that will lead to still more warming and more methane release.

## *CFCs AND THE GREENHOUSE EFFECT*

The CFCs also play an important role in the greenhouse process (Chapter 16). Although present in much smaller quantities, the impact is large. Because of the high amount of carbon dioxide already in the atmosphere, it is effectively blocking earth radiation. However, additional chlorofluorocarbon atoms added to the atmosphere are 10,000 times more effective at absorbing infrared radiation than are additional molecules of carbon dioxide. CFCs probably account for 25 percent of the increase in temperature. A quadrupling of the chlorofluorocarbons might produce an increase in temperature of from 0.5 to 1.0°C (1 to 2°F).

## *THERMAL POLLUTION*

Thermal pollution resulting from increasing human use of energy and the inevitable discharge of waste heat into the atmosphere or ocean is another source of heat. Although it is not yet significant on a global scale, this heat source may become appreciable by the middle of the next century. If future energy production is concentrated in large nuclear power parks, the natural heat balance may be upset even sooner.

The addition of carbon dioxide to the atmosphere is not the only potential anthropogenic cause of climatic warming. Thermal pollution resulting from increasing use of energy and the inevitable discharge of waste heat into either the atmosphere or ocean is another source of atmospheric heat. Although it is not yet significant on the global scale, this heat source may become an appreciable fraction (1 percent or more) of the effective solar radiation absorbed at the earth's surface by the middle of the next century. If future energy generation is concentrated into large nuclear power parks, the natural heat balance over considerable areas may be upset long before that time.

## MODEL PROJECTIONS OF TEMPERATURE CHANGE

One set of projections forecasts an increase in global temperature under a "business-as-usual" scenario. It is assumed that no controls are placed on greenhouse gas emissions. Under these conditions temperatures will be about 1°C above present levels (2°C above preindustrial levels) by 2025. Before the end of the twenty-first century, global mean temperature will be 3°C above the current level.

The main uncertainties in the projected temperature increase are related to a number of factors. First, there is still an incomplete understanding of the nature of sinks for carbon dioxide while, as may be anticipated, future additions to the atmosphere will depend largely on the actions and policies of governments in relation to fossil-fuel use. Second, the role of clouds in a greenhouse warmed climate is not yet fully understood. Whether clouds will be more efficient reflectors of shortwave radiation or absorbers of longwave radiation is not yet clear. Third, the role of the oceans, which influence both the pattern and timing of climatic change, has yet to be fully determined. There may be other carbon sinks as well.

## GLOBAL WARMING AND CLIMATIC CHANGE

To many it would seem that an increase in a fraction of a degree of temperature is of little consequence in the scheme of climate where day-to-day changes are in terms of tens of degrees. But change in the mean global temperature is but an indicator of the many climatic systems that occur over the globe; of major significance is that the change would lead to a modification of temperature gradients over the earth's surface, and this in turn would lead to a modification of the general circulation pattern. Changes in precipitation and storm tracks are but some of the related consequences.

Almost all of the debate concerns global warming as a result of addition of greenhouse gases to the atmosphere. There has been controversy, with some scientists taking a view that the case for such warming and associated change is overstated. Nonetheless, there is a general consensus that global warming is real, with the major areas of disagreement being the interpretation of the rate at which it is occurring and the extent of the impacts of that increase.

Perhaps the best view of the controversy is provided by the final report of the working group of the Intergovernmental Panel on Climatic Change (IPCC), which is sponsored jointly by the World Meteorological Organization and the United Nations Environment Program. The report was based upon the scientific assessment of climatic change by several hundred scientists in 25 countries to provide perhaps the most comprehensive view of opinions on the subject.

The scientists participating in the study are confident that carbon dioxide has been responsible for over half of the enhanced greenhouse effect in the past, and is likely to remain so in the future. The atmospheric concentrations of long-lived gases (carbon dioxide, nitrous oxide, and the CFCs) adjust only slowly to changes in emissions. Continued emission of these gases at present rates would commit the atmosphere to increased concentrations for centuries to come. To stabilize concentrations at their present level, emission of long-lived gases would need to be decreased by 60 percent.

## SUMMARY

There is a natural storage process operating in the atmosphere which raises the temperature at Earth's surface from below freezing to about 15°C. This storage process is the greenhouse effect. It results from some atmospheric gasses being fairly transparent to solar radiation but more opaque to Earth radiation. This results in energy flowing back and forth between the surface and the lower atmosphere.

Human industrial activity is now increasing the amount of some of the gases that are important in the greenhouse effect. Global temperatures have increased since the middle of the nineteenth century. Some scientists believe that this increase in temperature is due to the increase in the emission of greenhouse gases such as $CO_2$ and methane. Whether humans are forcing the temperature of the planet upward is not clear. Only time will provide the answer.

## BIBLIOGRAPHY

Bird, E. C. F. 1993. *Submerging Coasts: The Effects of a Rising Sea Level on Coastal Environments*. New York: Wiley.

Canning, D. J. 1991. Global climate change and sea level rise. *Currents*, 10:8–12.

Friis-Christensen, E., and K. Lassen. 1991. Length of the solar cycle, an indicator of solar activity closely associated with climate. *Science*, 254:698–700.

Warrick, R. A., E. M. Barrow, and T. M. Wigley. 1993. *Climate and Sea Level Change: Observations, Projections and Implications*. New York: Cambridge University Press.

Wuebbles, D. J. 1991. *A Primer on Greenhouse Effect Gases*. Chelsea, Mich.: Lewis Publishers.

chapter 19

# Potential Impact of Global Warming

***John E. Oliver***

***CHAPTER SUMMARY***



A change in temperature of 1.5 to 4.5°C in a century or two is unprecedented in the recent history of Earth. A 2°C rise in temperature would move the earth to a climatic position similar to that which existed during the climatic optimum 6000 years ago. Changes in global mean temperature of 2°C or more would have a significant impact on world society. The stress of adjustment would be phenomenal. A change of this size would affect nearly all aspects of life. It would certainly affect the production of food. Since agriculture began on earth, the mean temperature has not varied more than 1°C from the mean. Forecasts are for global temperatures to rise and continue to rise over the next century (Figure 19.1). If this happens, many other physical and economic changes will follow.

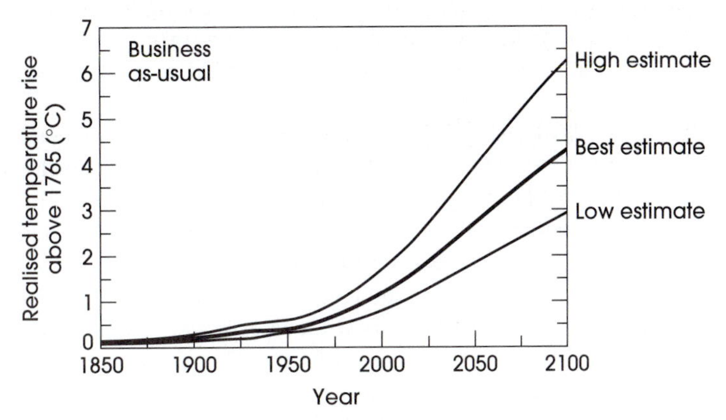


**Figure 19.1** Simulation of the increase in global mean temperature from 1850 to 1990 due to observed increases in greenhouse gases and predictions of the rise between 1990 and 2100 resulting from "business as usual" emissions. (From J. T. Houghton, G. J. Jenkins, and J. J. Ephraums, eds., *Climate Change: The IPCC Scientific Assessment,* Cambridge University Press for the Intergovernmental Panel on Climatic Change, New York, 1990.)

## GLOBAL WARMING AND SEA-LEVEL CHANGE

The processes that determine sea level are many and complex. Changes in the ocean level can occur if the volume of water in the ocean changes. Changes can also occur if the volume of the ocean basins changes. Most changes in the ocean basins require many years. The accumulation of sediments on the ocean floor and adding magma to the oceans floor are examples. Rapid changes result primarily from ice alternately forming and melting, and to a lesser extent from thermal expansion and contraction of sea water. At any specific coastal location it is very difficult to measure sea-level change. Both tectonic and climatic changes determine the relative height of sea level.

The Pleistocene Period provides much information about temperature change. During the Pleistocene there was rapid growth and melting of ice. The height of the ocean is the net record of melting of the ice sheets. As ice on the continents alternately grew and melted, sea level fell and then rose. During periods of warmer climate sea level rose. Glaciers melted and the ocean water expanded from the additional heat. The fall and rise of sea level during the Pleistocene is an important guide to glacial and interglacial periods. Submergence and emergence of coastal areas and marine terraces point to the amount of water stored as ice. With appropriate dating methods, such evidence provides a guide to glacial advance and retreat.

Analysis of various kinds of data shows that the last interglacial occurred some 120,000 years ago. On Greenland the ice sheet melted away enough to expose the land beneath in some areas. The ice sheet was both smaller in area and volume. Coral fossils from the last interglacial are about 6 meters (20 feet) above present sea level. This shows that sea level was some 6 meters (20 feet) higher during the last interglacial than now. If the remaining ice on Greenland were to melt, sea level would rise at least another 6 meters (20 feet).

The end of the last glacial advance was about 21,000 years ago. Some 5 percent of all planetary water was in the form of ice. This amounted to 58 million cubic kilometers of water withdrawn from the ocean. There was a staggering 50 million cubic kilometers (12 million cubic miles) of water stored in the northern hemisphere ice sheets alone. This lowered sea level by a large amount. Evidence from the Gulf of Mexico suggests that the ocean was between 125 and 140 meters below its present level.

At the height of the last glacial advance continental margins were very different from those of today. Once the ice sheets began to melt, sea level rose rapidly, if not steadily, 50 to 100 meters by 11,500 years ago (Figure 19.2). The level stabilized for a time, reflecting the cooler conditions around the North Atlantic Ocean. Once the climate continued to warm, sea level rose again. It has been rising since then. Over the past 6000 years, sea level has been within about 1 meter of its present level. In the past century sea level has risen between 0.1 and 0.2 meter. Over the past 100 years sea level has been rising at an average rate of between 1 and 2 millimeters (0.04 to 0.08 inch) per year. This has been concurrent with the melting of glaciers in midlatitudes.

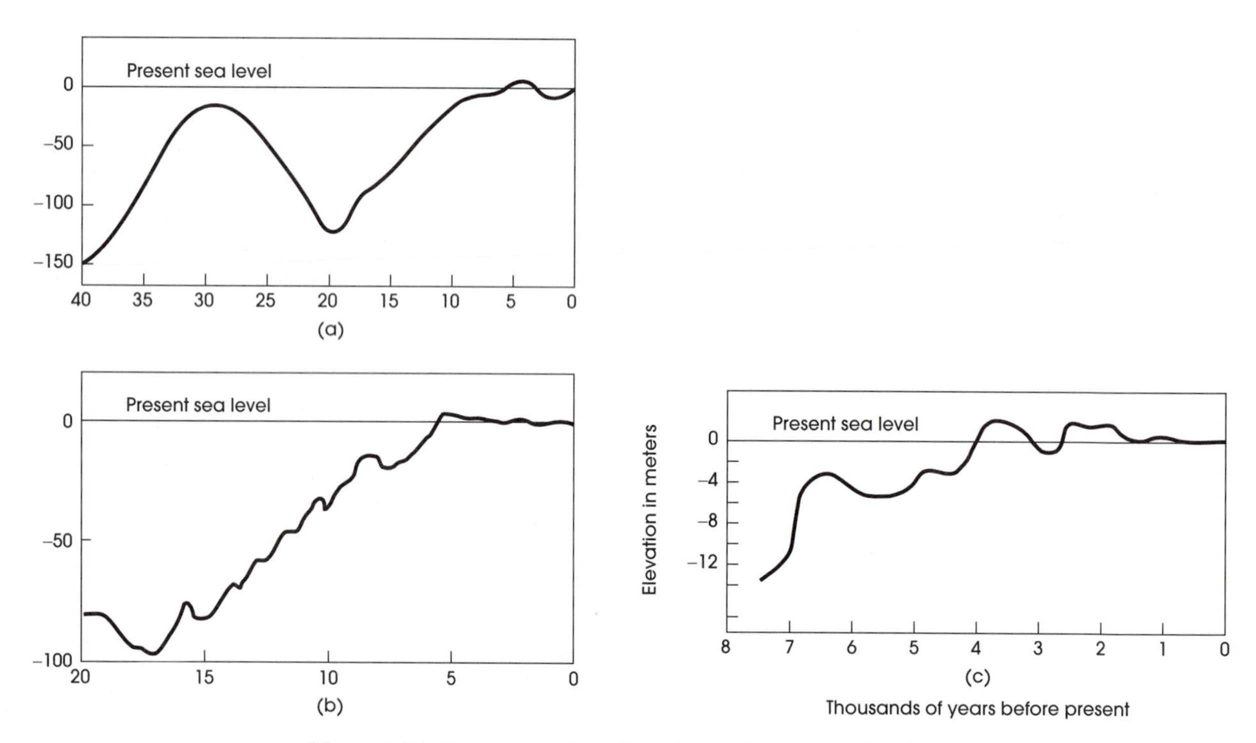


**Figure 19.2** Past and projected sea-level changes. (a) Sea-level change over the past 40,000 years. Sea level reflects the change in the amount of ice as the ice sheets changed in volume. (b) Sea-level change from the time of the Wisconsin glacial maximum to the present. (c) Change in sea level over the past 7000 years.

## *FUTURE SEA-LEVEL CHANGE*

If global warming takes place, sea level will rise due to two different processes. Warmer temperatures cause sea level to rise due to the thermal expansion of seawater. The effect of global warming will be to warm the water in the ocean and cause it to expand. Computer models suggest that this expansion may cause mean sea level to rise 40 to 80 millimeters by the year 2025 (Figure 19.3). The basis for this is the estimate that world temperatures would rise from between 0.6 and 1.0°C by the year 2025.

Water from melting mountain glaciers and the ice sheets of Greenland and the Antarctica would also add water to the ocean. Sea level can also rise as a result of the subsidence of coastal land areas. Thus the causes of sea-level rise, the absolute amount of sea-level rise, and the rate and timing of the rise are uncertain.

The melting of glacial ice is most likely to be the reason for sea-level rise in the coming centuries. A temperature increase between 0.6 and 1.0°C (1.2 to 1.8°F) may cause mean sea level to rise 40 to 80 millimeters (1.6 to 3.2 inches) by the year 2025. The National Research Council (1983) estimated a sea-level rise of 200 to 700 millimeters (8 to 28 inches) in the next century. If there is a rapid melting of sea ice, sea-level rise could increase to 1 or 2 meters ( 3 to 6.5 feet) in the next several centuries. There is a real hazard of rapid sea-level rise if mean global temperatures rise 2°C (4°F) or more.

There are two major geographical areas that may contribute to rising sea level. They are the ice on the southern part of Greenland and the West Antarctic ice sheet. The West Antarctic ice sheet has its base on the seafloor and hence would respond rapidly to changes in water and atmospheric temperatures. At present the summer temperature over the West Antarctic ice sheet averages −5°C (23°F). If summer temperatures should rise above freezing, very rapid melting would occur. This would result in a rise of sea level of 5 to 6 meters (16 to 20 feet).

Complete melting of the ice caps would change sea level some 70 meters (230 feet). However, the time it would take for the ice caps to respond to warming is quite long. It would take thousands of years for the ice caps to respond to a global warming of several degrees. This may be beyond the range of the possible effects of the $CO_2$ changes. It is not even certain if warming of the seas would cause the ice caps to grow or to shrink. The water may warm slightly, but not

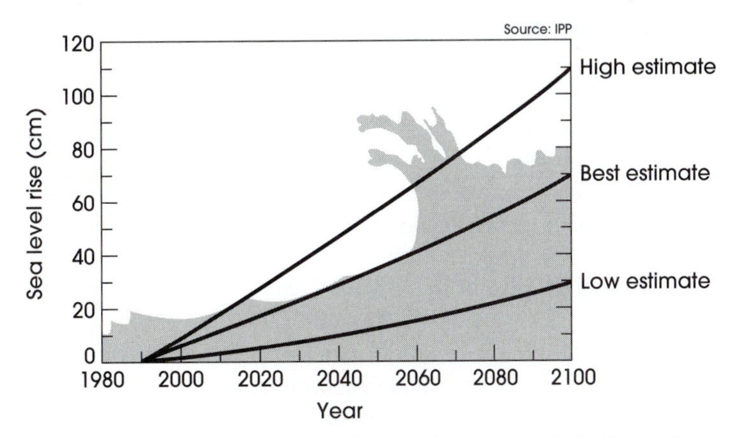


**Figure 19.3** Projected sea level depending on the extent of glacier melting and expansion of seawater. (From J. T. Houghton, G. J. Jenkins, and J. J. Ephraums, eds., Climate Change: The IPCC Scientific Assessment, Cambridge University Press for the Intergovernmental Panel on Climate Change, New York, 1990.)

enough to raise the temperature of the atmosphere over the ice above freezing. In this case, snowfall may increase and glaciers expand. If this happens, sea level will drop.

If the present Antarctic ice sheet were to melt, there would be enough additional water to cause a sea level rise of 60 meters (200 feet). The removal of ice from such a large land area would result in an upward movement of the land where the ice melted away. In some areas the rise of coastlines due to crustal decompression from the last ice age is taking place faster than sea level is rising. The net effect is one of apparent lower sea level. The ocean floors would sink further into the crust under the weight of additional water. Even with these adjustments, melting of the Antarctic ice would still cause a rise of 40 meters (135 feet). This would be enough to flood most of the world's major sea ports.

## *ECONOMIC IMPACT OF RISING SEA LEVEL*

Over half of the human population lives within 100 kilometers of the sea. Most of this population lives in urban areas that serve as seaports. A measurable rise in sea level will have a severe economic impact on low-lying coastal areas and islands. A rapid rise in sea-level of several meters would flood large sections of the earth's coastal plains including many of the world's great cities. The small rise in sea level now taking place is already causing major problems in some areas. London, England, Venice, Italy and the country of The Netherlands, are taking remedial measures to reduce the risk of flooding from the sea.

The horizontal effect of rising sea level can be greater than the actual rise would indicate. As sea level rises erosion increases along coastlines. This is particularly the case along flat coastlines. On emergent coastlines, a rise of 1 centimeter can increase the rate of beach erosion inland by 1 meter. Rising sealevel pushes saltwater inland. A 10 millimeter rise in sea level would push the boundary between salt water and fresh water 1 kilometer upstream along coastal rivers. Saltwater would also displace fresh groundwater for a substantial distance inland. Oceanic islands would be severely affected. Costs of keeping out the sea may be prohibitive and movement inland impossible. In these cases the only solution may be to move people and their institutions to some other location.

## *GLOBAL WARMING AND AGRICULTURE*

Agriculture is a major element in the U.S. economy. It contributed 17.5 percent of the gross national product in 1985. In that year farm assets totaled $771 billion. The Dust Bowl years showed that crop production is sensitive to climate, soils, and management methods. During the 1930s, wheat and corn yields dropped by up to 50 percent. More recently, during the drought of 1988, corn yields declined about 40 percent.

### Direct Effects of Increased $CO_2$

The probable effects of the increase in carbon dioxide on the biosphere vary. It is not normally toxic, as humans can breath concentrations of up to 15 times present levels before lethal results occur. Carbon dioxide is, however, a growth stimulant to green plants. Plants grow bigger and faster, contain more vitamin C and sugar, and are more disease resistant when grown in atmospheres with high concentrations of $CO_2$. Experiments show that some vegetables may increase in weight as much as 800 percent of normal in carbon dioxide–rich environments. Thus any increase in atmospheric carbon dioxide will be partly offset by increased use by the earth's biomass.

### Effects of Climate Shift

There will undoubtedly be major shifts in the general circulation if the atmosphere warms. There will be also be changes in the length of the growing season in midlatitudes, and there will be changes in precipitation patterns. What these changes will be is not known. We cannot predict what the net effect of these changes will be on world agriculture. Some changes will be offsetting. An increased length of growing season may be offset by lower rainfall.

Climatic changes due to atmospheric warming should not affect agriculture in the United States before the year 2000. Computer models show the growing season in the northern United States will increase by about 10 days and there will be more frequent droughts.

Crop yields could be reduced, although the combined effects of climate and $CO_2$ would depend on the severity of climate change. In most regions of the United States, climate change alone could reduce dryland yields of corn, wheat, and soybeans. Site-to-site losses would range from negligible amounts to 80 percent. These decreases would be primarily the result of higher temperatures, which would cause considerable heat and water stress in the plants. In southern areas where heat stress is already a problem, yields would drop. In areas where rainfall decreases, crop yields could decline. In response to the shift in relative yields, grain crop acreage in Appalachia, the southeast, and the southern Great Plains could decrease.

The combined effects of climate change and increased $CO_2$ may increase yields in some areas. This would be the case in northern areas or in areas where rainfall is abundant. There would be warmer temperatures and a longer growing season. Acreage in the northern Great Lakes states, the northern Great Plains, and the Pacific northwest could increase.

A change in agriculture would affect not only the livelihood of farmers but also agricultural infrastructure and other support services. Even under scenarios of extreme climate change, the production capacity of U.S. agriculture is estimated to be adequate to meet domestic needs. A decline in crop production would reduce exports, which could have serious implications for food-importing nations.

### Technological Response by Farmers

Improvements in crop yields could offset some negative effects of climatic change. In many regions, the demand for irrigation is likely to increase as a result of higher prices for agricultural commodities. Farmers may also switch to more heat- and drought-resistant crop varieties, plant two crops during a growing season, and plant and harvest earlier. Whether these adjustments would counter climate change depends on the severity of the climate change.

## *CHANGES IN NORTH AMERICAN FORESTS*

Forests occupy one-third of the land area of the United States. Temperature and precipitation are of major importance in the distribution of forests. Climate change could move the southern boundary of hemlock and sugar maple northward by 600 to 700 kilometers (about 400 miles). Migration of the forests will lag behind shifts in climate zones. The northern boundary would move only as fast as the rate of migration of forests. If there is a migration rate of 100 kilometers (60 miles) per century, or double the known historic rate, the geographical area of forests would drop. This is because the southern boundary may advance more quickly than the northern boundary.

Climate change may significantly alter forest composition and reduce the area of healthy forests. Seedlings of trees now growing in southeastern areas would not grow because of higher temperatures and dry soil conditions. In central Michigan, grasslands might replace forests now dominated by sugar maple and oak.

These analyses did not consider the introduction of species from areas south of these regions. In northern Minnesota the mixed boreal and northern hardwood forests could become entirely northern hardwoods. Some areas might experience a decline in productivity, while others might have an increase in productivity. The process of change in species composition would probably continue for centuries.

Other factors will also influence forests. Health of forests is not determined by climate change alone. The drier soils expected to accompany climate change could lead to more fires. Warmer climates may cause changes in forest pests and pathogens. Changes in air pollution levels could reduce the resilience of forests. Continued depletion of stratospheric ozone would stress forests further.

## *EFFECTS ON BIODIVERSITY*

Historic climate changes, such as the ice ages, have led to extinction of many species. More recently, human activities have accelerated the rate of species extinction. Climate warming would probably lead to an even greater loss of species. Natural ecosystems may suffer in negative ways and in ways that we cannot predict.

Forests, other plants, and animals may have difficulty migrating as fast as climate zones migrate. Species populations may decrease or become extinct. The presence of urban areas, agricultural lands, and roads would restrict habitats and block migration of some species. These obstacles may make it harder for plants and wildlife to survive future climate changes.

Some species may benefit from climate change. This could result from increases in habitat size or reduction in population of competitors.

## *EFFECTS ON AQUATIC SYSTEMS*

Fresh water fish populations may grow in some areas and decline in others. Fish in large water bodies such as the Great Lakes may grow faster and may migrate to new habitats. Increased amounts of plankton could provide more forage for fish. However, higher temperatures may lead to more aquatic growth, such as algal blooms. Decreased vertical mixing in lakes would deplete oxygen levels in shallow areas of the Great Lakes. This would make some lakes, such as Lake Erie, less habitable for fish. Fish in small lakes and streams may be unable to adapt to changing temperatures. In other cases their habitats may simply disappear.

The full impact on marine species is not known. The loss of coastal wetlands could certainly reduce fish populations, especially shellfish. Increased salinity in estuaries could reduce the abundance of freshwater species but could increase the presence of marine species.

## *EFFECTS ON THE HYDROLOGIC CYCLE*

Global precipitation is likely to increase. However, it is not known how regional rainfall patterns will change. Some regions may have more rainfall, while others may have less. Furthermore, higher temperatures would probably increase evaporation. These changes would probably create new stresses for many water management systems.

Results of hydrology studies show that it is possible in some regions to identify the direction of change in water supplies and quality. For example, in California, higher temperatures would reduce the snowpack and cause earlier melting. Earlier runoff from mountains could increase winter flooding and reduce deliveries to users. In the Great Lakes, reduced snowpack, combined with potentially higher evaporation, could lower lake levels.

Changes in water supply could significantly affect water quality. Where river flow and lake levels decline, such as in the Great Lakes, there would be less water to dilute pollutants. Where more water is available, water quality may improve. Higher temperatures may enhance thermal stratification in some lakes and increase algal production, degrading water quality. Changes in the amount of runoff, leaching from farms, and increases in irrigation could affect water quality in many areas.

## EFFECTS ON DEMAND FOR ELECTRIC POWER

Changes in climate, technology, and economic growth will alter the demand for electricity. The principal climate-sensitive electricity end uses are space heating and cooling and, to a lesser degree, water heating and refrigeration. These uses of electricity may account for up to a third of total sales for some utilities.

Global warming would increase annual demand for electricity and total generating capacity requirements in the United States. The demand for electricity for summer cooling would increase. The demand for electricity for winter heating would decrease. Annual electricity generation in 2055 is estimated to be 4 to 6 percent greater than without climate change. The annual costs of meeting the increase, assuming no change in technology or efficiency, is estimated to be $33 to $73 billion (in 1986 dollars). States along the northern tier of the United States could have net reductions in annual demand of up to 5 percent. Decreased heating demand would exceed increased demand for air conditioning. In the south, where heating needs are already low, net demand was estimated to rise by 7 to 11 percent by 2055.

## EFFECTS ON AIR QUALITY

Air pollution caused by emissions from industrial and transportation sources is a problem in the United States. By reducing emissions over the last two decades, considerable progress has been made in improving air quality. Yet high temperatures in the summer of 1988 helped raise tropospheric ozone levels to all-time highs in many U.S. cities.

A rise in global temperatures would increase anthropogenic and natural emissions of hydrocarbons and emissions of sulfur and nitrogen oxides. The potential impact of increased emissions on air quality is uncertain. Higher temperatures would speed the reaction rates among chemicals in the atmosphere. This would cause higher ozone pollution in many urban areas. Higher temperatures would increase the length of the summer. This is a time of high air pollution levels. Preliminary analyses of a 4°C temperature increase in the San Francisco Bay area assuming no change in emissions from current levels, suggest that maximum ozone concentrations would increase by 20 percent. The area exceeding the National Ambient Air Quality Standards would almost double.

Studies of the southeast also show expansion of the areas violating the standards, but they show smaller changes in levels. It is likely that sulfur and nitrogen would oxidize more rapidly under higher temperatures. The ultimate effect on acid deposition is difficult to assess because changes in clouds, winds, and precipitation patterns are uncertain.

## EFFECTS ON HUMAN HEALTH

Weather affects contagious diseases such as influenza and pneumonia and allergic diseases such as asthma. Mortality rates, particularly for the elderly and the very ill, reflect frequency and severity of extreme temperatures. Changes in temperature and rainfall affect the life cycles of disease-carrying insects such as mosquitoes and ticks. Finally, increased air pollution can heighten the incidence and severity of such respiratory diseases as emphysema and asthma.

Global warming may lead to changes in morbidity and increases in mortality, particularly for the elderly during the summer. If the frequency or intensity of climate extremes increases, mortality is likely to rise. If people acclimatize by using air conditioning, changing their work place habits, and altering the construction of their homes and cities, the impact on summer mortality rates may be reduced.

Changes in climate as well as in habitat may alter the regional prevalence of vector-borne diseases. For example, some forests may become grasslands, thereby modifying the incidence of vector-borne diseases. Changes in summer rainfall could alter the amount of ragweed growing on cultivated land, and changes in humidity may affect the incidence and severity of skin infections and infestations such as ringworm, candidiasis, and scabies. Increases in the persistence and level of air pollution episodes associated with climate change would have other harmful health effects.

## DIVERGING VIEWS OF THE IMPACT OF GLOBAL WARMING

Global warming is not a very important issue to most people in the world. There are much more pressing issues for individuals and families. Unemployment, crime, basic health care, and poverty are far more real problems. There are a number of groups of people most concerned about the potential of global warming. Among these are academics whose job it is to provide information about global warming and the planners that must deal with the problem.

A survey of scientists and economists published in 1994 (Nordhaus, 1994) revealed widely different views on the potential impact of global warming. The study shows that economists generally lack concern about global warming. They believe that the human species is so adaptable and technology is growing so fast that we could cope with the changes that would occur. They believe that since the changes would take place over a century or more, society could respond without much trauma. Most believe that if the temperature changes by 3°C by 2090 the impact will not be much.

Natural scientists, on the other hand, are much more concerned. Natural scientists estimated the probability of a high economic impact from global warming to be 20 to 30 times greater than did economists. Most studies of the economic impact of global warming focus on the industrial countries. Most scientists believe that the economic impact of global change will be greater in the developing countries than in the high-income countries.

There is little agreement about the impact of climatic change on different regions of the globe. Tropical regions would experience the least climatic change. Tropical regions are defined here as that part of the earth between 30°N and 30°S. Temperature would change the least, so atmospheric circulation and precipitation would change little. That climate would change the least in tropical regions is very important from the point of global economics. Most of the developing countries are in this zone. About half of the total world population lives in this zone. However, the region accounts for only about 16 percent of world production. Thus the countries that have relatively low per capita incomes will be affected the least. The major impact in the tropical regions would be changes in sea level. These changes would be important but less so than if they were coupled with climatic change.

## *SUMMARY*

Any change in climate affects world economic systems. If global temperatures change by 1°C or more, it will result in significant changes in the global environment. There will be changes in sea level, changes in the distribution of biomes, and changes in biodiversity. These changes will, in turn, result in changes in global economic systems. Changes in economic systems are accompanied by changes in social systems and widespread human stress. As the human population grows the number of people affected increases. Any unnecessary changes in the earth environment need to be avoided to reduce this human stress. "The perils of global warming, like those of ozone depletion, allow for little margin of error, since their impacts are far-reaching, multifaceted, and will last for many years to come. "By the time global warming can be measured, it will be too late to do much about it" (Committee on Earth and Environmetnal Sciences, 1990, p. 16 ).

## BIBLIOGRAPHY

ABRAHAMSON, D. E. ed. 1989. *The Challenge of Global Warming*. Natural Resources Defense Council. San Francisco: Island Press.

NORDHAUS, W. D. 1994. Expert opinion on climatic change. *American Scientist*, 82(1):45–51.

COHEN, S. J. 1990. Bringing the global warming issue closer to home: the challenge of regional impact studies. *Bulletin of the American Meteorological Society*, 71:520–526.

COMMITTEE ON EARTH AND ENVIRONMENTAL SCIENCES. 1990. *Our Changing Planet: The FY 1991 Research Plan of the U.S. Global Change Research Program*. Washington, D.C.: Government Printing Office.

EDGERTON, L. T. 1991. *The Rising Tide: Global Warming and World Sea Levels*. San Francisco: Island Press.

WYMAN, R. L. 1990. *Global Climate Change and Life on Earth*. New York: Chapman & Hall.

chapter 20

# Future Global Environmental Change and the Human Species

### CHAPTER SUMMARY



The key element in the future relationship between the human species and the global environment is the continued rapid growth of the human population. In 1968, Stanford biologist Paul Ehrlich published the book *The Population Bomb*. At the time he wrote the book, the human population was 3.5 billion. Based on present growth rates, the population is expected to reach 6.35 billion by the year 2000. About 90 percent of the additional growth will be in the developing countries, and about 80 percent of the world's population will be living in the developing countries.

The trend toward urbanization is predicted to continue. The percent of the population living in urban areas was 41 percent in 1975 and is expected to increase to 56 percent by the end of the century. This trend toward urbanization reverses a pattern of settlement that existed throughout the history of the species until the past 200 years. Agriculture was the means of livelihood throughout most of historic time. In the developed countries urbanization and industrialization took place coincidentally.

The third-world countries make decisions that will certainly add to the global increase in population. At the 1974 United Nations meetings in Bucharest and Rome, spokesmen for these nations asserted that they had no population problems. They defended two policy statements: (1) the hungry nations have the right to produce as many children as they please, and (2) others have the responsibility to feed them. The growing population is not because parents want children, but because they want *more* children. In Moslem countries, for example, the desired number of progeny per couple is "as many as God will send." This turns out to be an average of seven.

Family planning has not been successful in many countries. India has spent huge sums of family planning over many years and has accomplished little. In 1951 the population in that country grew by 3.6 million. In the 1970s it was growing at a rate of 16.2 million per year. Forecasts are for India to become the most populous country on Earth, surpassing even China.

Population growth in some areas of the world has risen well beyond the carrying capacity. The result is a deterioration in quality of life and degradation of the environment. In the developed world economic development also degrades the environment through increased resource exploitation. All technological developments from the early use of stone tools and fire to the invention of the CFCs cause some change in the environment.

## *FUTURE FOOD SUPPLY*

The problem of providing food for additional millions of people comes into sharp focus when one realizes that a large portion of the world does not now get adequate food. The majority of people in Latin America, Africa, and Asia do not get minimum protein and calories required to sustain a person in North America or Western Europe. No one knows how many people the earth can feed, as it depends on how much food can be produced and at what nutritional level the people are to be fed.

At the same time that the demand for food is growing by leaps and bounds, there are some problems arising which actually cause a reduction in the production of foodstuffs in some areas at least. One of these problems is the general downturn in grain production per unit area. From 1962 until 1972 the world trend was for yields per unit area to increase. It increased from 1.4 metric tons per hectare to 1.91 metric tons per hectare. From the peak in 1972 mean yields began to drop. Constraints of fertilizer, energy, land, and water are all beginning to contribute to decreases in yield.

### Marine Food Production

Prophecies of large increases in food production from the sea seem to be far from feasible. In the best fishing areas, the seas are indeed productive, but such areas are limited in extent. At mid-century the world fishing industry was brimming with optimism. Seemingly inexhaustible supplies of fish could be netted as fast as fishing technology improved and capital investment expanded. The optimism was well founded. The world fish catch expanded from 22 million tons in 1950 to 70 million tons in 1970. Then suddenly, with little warning, the world fish catch declined. Overfishing is now affecting a wide variety of fisheries in both the Atlantic and Pacific Oceans.

### Terrestrial Food Production

A commonly suggested answer to the increased need for food is to bring more land into production. The tropical rain forests are the only region left to exploit. These forest regions are in precarious balance with the environment under the best of conditions. Soils are poor in mineral nutrients and are not productive under European agricultural systems. The rain forests, deserts, and arctic regions are not going to become productive under present technology. Repeated attempts at introducing European-style agriculture have stumbled. Brazil has increased agricultural production in some commodities. They are the world's leading producer of sugarcane and the third largest producer of soybeans. Much of this production has not come from the conversion of virgin lands to cropland but from changing from local market crops to export crops.

At present, more irrigated land fails each year than is placed into production. There are many problems associated with irrigation in desert regions. The lack of available water may be the principal constraint on future efforts to expand world food output. The expansion of irrigated land does not hold great promise. The best floodplain sites in deserts have long since been used for irrigated agriculture. From 1950 to 1970 there was a great expansion in irrigated land in China, India, and other developing countries. The total irrigated area is now expanding at less than 1 percent per year.

The momentum toward famine is at this time so great that there seems to be no way of halting it. The only hopeful possibility is to reduce the dimensions of the coming disaster. It makes no difference how much food the world produces if it increases the population faster. Some nations are chronically on the brink of famine because their populations have grown beyond the carrying capacity of the land. The population of the developing countries is increasing at nearly 70 million persons per year.

### Sustainable Agriculture

Since the 1980s, American farmers have been actively fighting soil erosion by trying new approaches to farming called sustainable agriculture. At one time this term was used for organic farming that used no synthetic chemicals. Today the definition means a wide range of environmentally sound practices. Many of these practices relate to reducing erosion.

One method is to "farm ugly," using farming methods consisting of leaving on the land clods of dirt and ragged stalks left over from last year's harvest. The clean fields of the past were suited to efficient use of farm machinery but encouraged erosion. Farmers in South Dakota are planting wheat, sunflowers, soybeans, and corn in fields littered by the debris from previous harvests. The trash serves the purpose of providing nutrients for the soil and allowing the water to soak in. Fields containing leftover stubble and planted to wheat produce higher than average yields.

Another deterrent to erosion is the use of grass buffer strips. The grass seed provides a profit and the grass serves to capture runoff and sediment. When used as set-aside land, grass buffer strips can qualify for federal cost sharing, which encourages farmers to fit them into federal commodity and conservation programs.

Indians in Central America interplanted corn, squash, and beans in precolonial times. Some American farmers are experimenting with modern variations on the technique by planting rows of corn between rows of sugar beets. This offers the beets shelter from high winds and corn more access to sunlight. The result has been higher yields of both corn and beets. Farmers are beginning to adapt methods of sustainable farming because of economic necessity. Chemicals are expensive and reduce organic content.

Population growth decreases world food supply by reducing arable land. The amount of arable land available is essentially fixed. Each additional person added to Earth's population requires about 0.08 hectare of land for living space. At the time the original 13 colonies were established in the United States there were 242.9 hectares of land available per capita. By the year 2000 this will be down to 3.2 hectares per capita. If the arable land remained constant, the total number of people that could be sustained could be determined. However, large amounts of land are taken out of production by the expansion of cities and construction of highways, powerlines, and pipelines. By the year 2000 an additional 8 million hectares will have been withdrawn for these purposes in the United States. This is an amount of land equal in area to the combined states of Vermont, New Hampshire, Connecticut, and Rhode Island.

## *OPTIONS FOR THE FUTURE*

It is very clear that there are options open to our species. As the first Club of Rome stated, "Decisions are being made every day, in every part of the world, that will affect the physical, economic, and social conditions of the world system for decades to come (Meadows et al., 1972, p. 28). It is important to all of us what these decisions are.

## *OPTION 1: BUSINESS AS USUAL*

Decision can be made to opt for the present pattern of population and industrial growth. One of the goals of conservation is "the greatest good for the greatest number of people." Implicit in this philosophy is the assumption that the number of people can continue to increase and we can continue to increase the production of goods and services per person. If the resources available to the human species are limited to this planet, there is a finite fund of all resources, including energy and fresh water and agricultural land. If there is a limit on our available resources, it should be evident that as population grows, the amount of resources that can be allocated per person must decrease. There is a limit to the number of people that can be supported at any economic level. It does not appear that there are enough resources available to support the world population at anywhere near the standards of the average person living in the developed countries.

### Environmental Consequences of Present Trends

At this time it appears that as the world population continues to grow, the impact on the global environment is going to be large. It does not seem possible to avoid accelerated alteration and destruction of the human habitat. The following forms of degradation will take place, but for how long and to what extent is not known.

1. Deforestation will continue unabated.
2. Global biodiversity will continue to decline.
3. Overgrazing will increase on a worldwide scale.
4. Soil erosion will worsen.
5. There will be more pollution as technology increases in extent and variety.
6. There will be general environmental degradation in forms not yet recognized.

The form of degradation will be different from society to society, depending on the state of development of the culture involved.

### Socioeconomic Consequences of Present Trends

1. There will be more crowding as the population grows and becomes concentrated in larger and larger cities.
2. There will be fewer freedoms for the individual.
3. There will be less personal privacy.
4. There will be increased personal tensions.
5. There will be more violence.
6. There will be more pollution-related problems.
   a. More odor pollution
   b. More sound pollution
   c. More visual pollution
   d. More new product problems

## *OPTION 2: SUSTAINABLE DEVELOPMENT*

*Carrying capacity* is the limit of a given population the environment can sustain without damage to the environment. The ultimate limit to carrying capacity on the planet is photosynthesis. This is the limit of the ability of green plants, algae, and bacteria to convert solar energy to living organic matter. The carrying capacity of the planet for the human species has limits. When population increases and economic growth exceeds a threshold level, environmental degradation takes place. This degradation is now almost universal on the planet. It varies in form from place to place and varies in degree, but it is there.

There is growing support for a global policy of sustainable development. The major premise is that there must be a balance among population growth, economic growth, and the carrying capacity of planet. Sustainable development must provide a balance between the needs and desires of the present generation and yet protect the carrying capacity of planet for future generations. This is the point where global environmental change and global change merge into a single interactive system. Planning for the future cannot consider the environment alone or human wants alone. Economic growth and environmental stability are inextricably linked.

Sustainable development recognizes that there are limits to human population growth and economic growth. The recognition of the need to establish this balance is not new. British economist Thomas Malthus wrote in 1798 on the principles of population growth. He wrote that humans will increase their numbers beyond the resources available to sustain these numbers. At that time, disease, famine, and war will reduce the population to a size that the system can support.

There are a group of academics, mainly social scientists and economists, who argue that there is no carrying capacity at all. There is no limit to population growth. They base their argument on the fact that all projections of shortages of food, energy, and other resources have failed to materialize. We have always been able to find substitutes for items in short supply. We have been able to increase food supply not only to keep up but have been able to increase the percentage of the world population that has a satisfactory level of nutrition. They continually point to the Malthusian doctrine as being not only scientifically but morally wrong.

That there are regional limits to growth has been known for centuries. *The Limits to Growth* was perhaps the first major statement that recognized the problem at the global scale. There is an upper limit to the sum of population growth, food production, and economic productivity. The better the quality of life is going to be in the future, and the better the diet people are to have, the lower the total global population will need to be.

What the sustainable population may be in which world resources are distributed equally is not known. Paul Ehrlich recognizes that there is no fixed number of people attached to the carrying capacity. We can make choices about the level of food intake and the economic level at which we wish to live. The greater per capita income and the higher the level of nutrition we choose, the lower the number of people there will be under the carrying capacity.

Present population is close to 6 billion and is projected to grow as high as 20 billion. In 1994, David Pimentel, a professor at Cornell University, reported on a computer model that projected a sustainable global population of 1 to 3 billion. For the United States he suggests a sustainable population of 200 million. The basis for his argument is that there are trade-offs in future development. We can increase one resource but must accept reduction in others. The system is an interdependent whole and the limit to the entire system has been reached.

Projected population numbers for a sustainable world society and for the United States are well below those that already exist. There are many choices to be made, depending on the quality of life we wish to establish. The higher the quality of life, the lower the sustainable population.

Another factor that is beginning to emerge is that the productivity of the entire system may be declining. There is considerable evidence that productivity of the world ocean is declining. Current population growth demands that there be millions of hectares of new agricultural land each year. This demand exists in the face of some 1 million hectares of land being abandoned to agriculture each year.

If sustainable development is to take place, it will be achieved by worldwide consensus and individual action. Thomas Malone (1991, p. 14) states that there are two requirements for sustainable development: (1) a change in our way of thinking—metanoia, and (2) transformation of the institutions by which human affairs are managed on a finite planet presently undergoing change of a global scale—change that is driven by human activity.

In all likelihood, the people of the world and the nations of the world will not accept the need for taking the appropriate steps to achieve a sustainable society in the near future. The basic underlying concept is that the earth is a single holistic system that includes but is not possessed by the humans species. It will demand the recognition by the scientific community that input from all disciplines is essential to effective planning. There is no simple engineering solution or political solution. There must be an acceptance of the interdependence of knowledge and process.

Political interdependence must also be recognized. We live in a single world environment. Global environmental change affects all people and all nations. There is now a single global economic system. The United States depends upon imports for the majority of its needs for aluminum, chromium, manganese, nickel, tin, zinc, iron, lead, and tungsten. By the year 2000 it will also depend on imports for more than half of its copper, potassium, and sulfur. Every nation is economically dependent on almost every other. Of the 13 basic raw materials used by industry, only phosphorus is in plentiful supply. The United States is now tied into a worldwide economic system from which we could not and would not want to extricate ourselves. This must be accepted by all nations before major progress can be made toward a society in equilibrium with the planetary environment. This is not now the case. The nations of the world still act as independent entities politically, economically, environmentally, and socially.

At current growth rates, more than 90 percent of the entire increase in world population will occur in the lesser developed countries, which have a per capita GNP about one-twentieth that of the richer nations. The growth rates in the developing countries are still much higher than they were in Western Europe when they passed through the rapid growth phase of the demographic transition, and this will continue to shift a greater proportion of the world's population into the regions of Asia, Africa, and Latin America (Table 20.1).

With more than a third of the people in these poor countries under 15 years of age, there is already a tremendous potential for future population growth. Even if the number of people using birth control was doubled immediately, population would continue to grow by large numbers for at least another 15 years because of the enormous increase in the number of people who will be entering the childbearing years during this time. Thus reaching 8 billion people will be delayed only a few years. Over the long term, of course, lower birthrates will result in a substantially lower population than will be the case if fertility remains near current levels. This must be the long-term goal, but results will not be apparent within only a few years.

On the positive side, modern means of fertility control and agricultural technology can now offer the possibilities of being able to lower birthrates and to increase food production at a faster pace than took place in Europe during its demographic transition. Moreover, the developed nations have financial resources that could perhaps be applied to solving population growth problems in the poor, developing countries.

## Regional Inequities in Resources and Quality of life

In planning for a sustainable world system it must be recognized that there are and will always be regional differences in resource availability. There are great inequities between regions in terms of economic growth and great differences between areas in terms of resources for future growth. Like resources, people are not distributed equally over the surface of the Earth. As discussed in Chapter 10, the regional distribution of people is always changing through varying birthrates, death rates, and migration. The only means by which nations can increase their population and standard of living at the same time is to increase economic activity to support them. While the developed countries use more and more of the world's resources to raise their quality of life, nearly 20 percent of the human population lives in abject poverty and near starvation. In the early 1990s the disparity between the richest and poorest countries was very large. The richest 20 percent of the human population consumes 70 percent of global economic output and the poorest 20 percent uses only 2 percent of the global economic output.

## Acceptance of Technology and Economic Growth

Other factors play a part in this widening gap as well. One such factor is the rate at which innovations are accepted. In the United States it took about 40 years for the acceptance of kindergarten as a part of public schools. It took some 18 years for the nation to accept driver's training as an

**TABLE 20.1** World Population Distribution by Region, 1650–2021 (Percent)

| | *Region* | | | | | |
|---|---|---|---|---|---|---|
| *Date* | *Asia* | *Africa* | *Latin America* | *U.S. and Canada* | *Europe* | *Oceania* |
| 1650 | 59 | 20 | 1 | <1 | 19 | <1 |
| 1800 | 66 | 10 | 2 | 1 | 20 | <1 |
| 1900 | 57 | 8 | 4 | 5 | 26 | <1 |
| 1940 | 55 | 8 | 6 | 6 | 24 | <1 |
| 1994 | 60 | 12 | 8 | 5 | 13 | <1 |
| 2021[a] | 60 | 18 | 8 | 4 | 9 | <1 |

[a]Projected.

integral part of public education, and more recently, five years of modern mathematics to be incorporated. The time it takes for the acceptance of innovations has steadily decreased. It took just two years for oral contraception to be widely adapted. In the developing countries, where technology is not an integral part of everyday life, the acceptance rate is much slower than that of the developed countries. Since economic growth is dependent on technology and since the growth rate is exponential, the developing countries will get further and further behind the developed countries.

This large imbalance in resource allocation and the poverty in some countries have the potential for initiating social tensions. The differential growth rates create pressures against political boundaries and create tensions in the increasing competition for the scarce resources of Earth. These differential growth rates provide the impetus for voluntary migrations and are often the motive in forced expulsion or the flight of refugees. A large part of migration of people today results from this inequity. This type of forced expulsion has been common in Africa in recent years. Tribes that have moved across boundaries have been driven back forcefully. In some cases, minority peoples have been expelled forcefully after many generations of living in a region.

Even in the rich countries such as the United States people are in the process of making some fundamental adjustments to higher costs of food, fuel, and land. Costs of goods and services are rising at a very high rate, more than the average middle-income family can keep up with. Most families now have two wage-earners. There are practical changes occurring in the way people live. There are smaller, less elaborate houses, fewer single-family dwellings, and more cluster housing.

### The Challenges Ahead

A tremendous challenge lies before all the citizens of the world to ensure that all people have freedom from hunger, want, and despair. This can be done by everyone working together:

1. To narrow the poverty/hunger gap both among nations and within each one by developing more effective and equitable economic development and social policies and by controlling population growth.
2. To lessen human impact on the environment by slowing population growth and eliminating excessive consumption and waste.
3. To use wisely the resources needed by each country so that no nation will be held hostage or suffer blackmail because of its dependence on another's resources as a consequence of one's own wasteful consumption.
4. To strengthen diplomatic attempts to reduce military confrontation in resolving political disputes that have arisen as a result of centuries of migrations, colonization, and arbitrary border designations.

Yet many parts of the world are beset by difficulties related to other aspects of the population, such as its distribution and migration patterns, its religious, ethnic, and racial composition within various areas, and its economic and cultural patterns based on heritage and geographical setting. Only when we become aware of the demographic forces that influence the myriad of seemingly insurmountable problems that face us in our constantly changing world will we begin to understand the issues of hunger, housing shortages, resource scarcity, lagging economic development, and political and social revolutions which pose a grave threat not only to certain regions but to all those in today's world.

## DEVELOPMENT OF THE U.S. GLOBAL CHANGE RESEARCH PROGRAM

The goal of the U.S. Global Change Research Program is to establish the scientific basis for national and international policymaking relating to natural and human-induced changes in the global earth system (Figure 20.1). To meet this goal, three major objectives have been established:

1. Establish an integrated, comprehensive long-term program for documenting the Earth system on a global scale.
2. Conduct a program of focused studies to improve our understanding of the physical, geological, chemical, biological, and social processes and trends on global and regional scales.
3. Develop integrated conceptual and predictive Earth system models.

## BENEFITS OF THE U.S. GLOBAL CHANGE RESEARCH PROGRAM

Specifically, the program is designed to achieve:

1. Improved abilities to predict the characteristics of global and regional changes that are needed to anticipate and plan for impacts on such important sectors as commerce, public safety, agriculture, energy, national security, natural resources, and international relations.
2. Enhanced abilities to separate natural changes from human-induced changes, thereby balancing regulatory needs with economic and social development and providing the ability to focus on those parts of the problem that are amenable to adaptation, mitigation, and human intervention.
3. Better knowledge of the connections between Earth system processes and economic, political, and social behavior.

The primary focus in the 1991 program was to establish the research required to provide the scientific information needed as input to current environmental policy discussion topics such as stratospheric ozone depletion and climatic change, including greenhouse warming. The scientific objectives of the research plan are to observe, understand, and ultimately to predict global change.

The program will initiate the development of NASA Earth Probes and Earth Observing System, key elements in the "Mission to Planet Earth," which will provide the focal point of an international satellite program for monitoring and understanding global changes.

## SUMMARY

There are two separate but interactive aspects to future global change. One is the natural component. To better predict future global change, we need to better understand changes that have occurred in the past and the processes that caused those changes. The second aspect of global change relates to human population growth and human activities. In the latter area, world society makes choices that will affect the future global system.

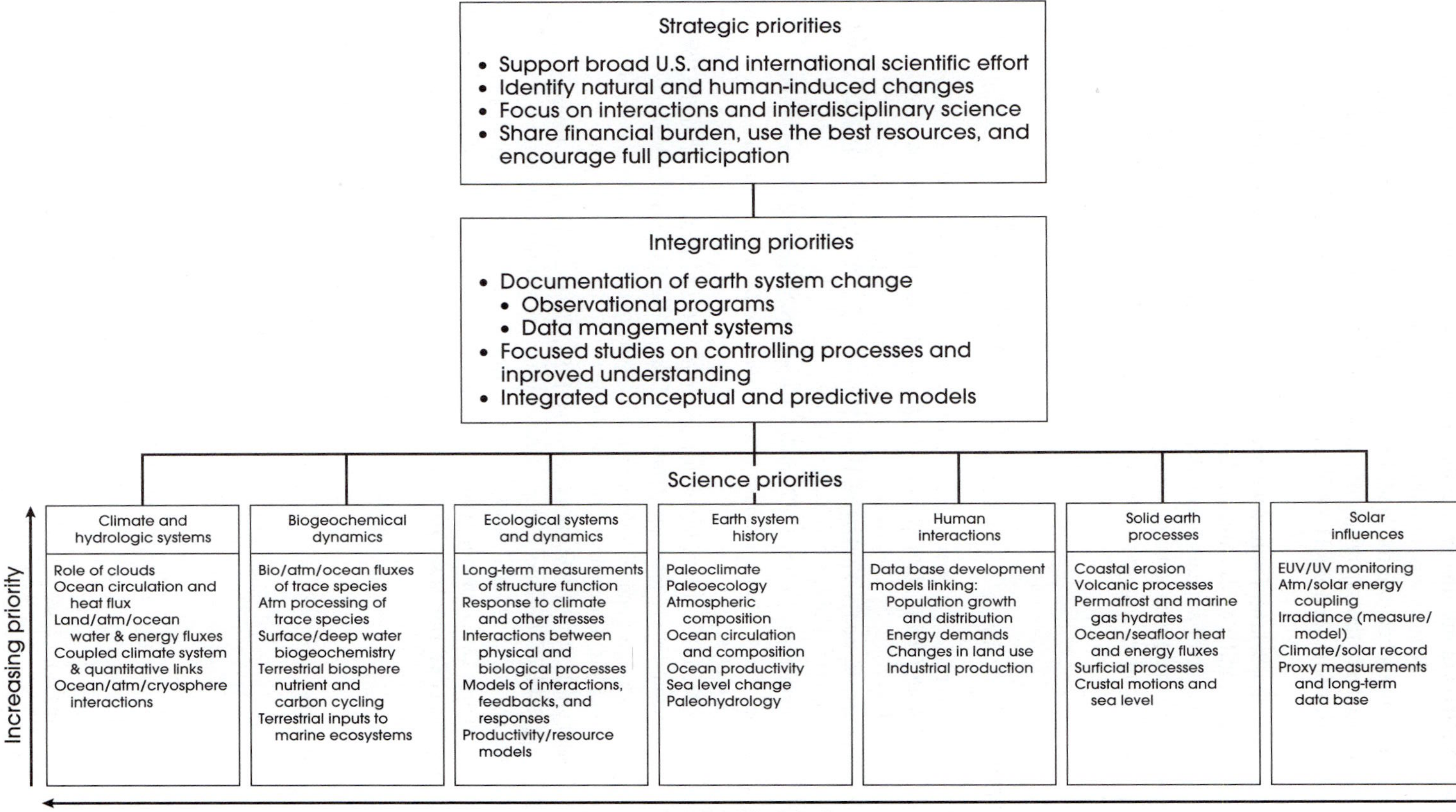


**Figure 20.1** Science priorities for the U.S. Global Change Program. From Committee on Earth and Environmental Sciences, *Our Changing Planet: The FY 1991 Research Plan of the the U.S. Global Change Research Program,* U.S. Government Printing Office, Washington, D.C., 1990, p. 13.)

It is extremely critical that decisions be made as soon as possible, for the longer the present trends continue, the greater the problems will be in bringing about a semblance of equilibrium, or a sustainable society, and the greater human suffering will be. Delays in decision making are also hazardous because the lead times for action and results on a global scale are very long. We certainly have the capability of making the right decisions, but whether we have the will or not remains to be seen. As H. G. Wells wrote in *The Time Machine* (1895, p. 79)

> It is a law of nature we overlook, that intellectual versatility is the compensation for change, danger, and trouble. An animal perfectly in harmony with its environment is a perfect mechanism. Nature never appeals to intelligence until habit and instinct are useless. There is no intelligence where there is no change and no need to change. Only those animals partake of intelligence that have to meet a huge variety of needs and dangers.

## BIBLIOGRAPHY

COMMITTEE ON EARTH AND ENVIRONMENTAL SCIENCES. 1990. *Our Changing Planet: The FY 1991 Research Plan of the U.S. Global Change Research Program.* Washington, D.C.: Government Printing Office.

EDBERG, R., and A. YABLOKOV. 1994. *Tomorrow Will Be Too Late: East Meets West on Global Ecology.* Tuscon, Ariz.: University of Arizona Press.

ERHRLICH, PAUL. 1968. *The Population Bomb.* New York: Ballentine Books.

GATES, D. M.. 1993. *Climate Change and Its Biological Consequences.* Sunderland, Mass.: Sinauer Associates.

MALONE, T. F. 1991. "Global Change in the 1990s: Imperatives for Action." In Corell, R. W. and P. A. Anderson. 1991. *Global Environmental Change.* Berlin: Springer-Verlag.

MATHER, J. R., and G. V. SDASYUK, EDS. 1991. *Global Change: Geographical Approaches,* (Vol. 1). Tuscon, Ariz,: University of Arizona Press.

MCCAY, B. J., and J. M. ACHESON, EDS. 1987. *The Question of the Commons: The Culture and Ecology of Communal Resources.* Tuscon, Ariz.: University of Arizona Press.

MEADOWS, D. H., D. L. MEADOWS, J. RANDERS, and W. W. BEHRENS. 1972. *The Limits to Growth.* Washington, D.C.: Potomac Associates, University Books.

Wells, H. G. 1895. *The Time Machine.* London: Publisher Unknown.

# Glossary

***Acid*:** Typically a water-soluble, sour compound that is capable of reacting with a base to form a salt; a hydrogen-containing molecule or ion able to give up a proton to a base or able to accept an unshared pair of electrons from a base.

***Acid rain*:** Rainfall with pH generally of less than about 5 on an annual basis owing to the presence of acidic substances of nitrogen and sulfur dissolved in the rain-water.

***Albedo*:** The reflectivity of the earth environment, generally measured in percentage of incoming radiation.

***Alluvium*:** Recent sediments deposited by streams.

***Anomaly*:** Deviation of temperature or precipitation in a given region over a specified period from the normal value for the same region.

***Anthropogenic*:** Of, relating to, or influenced by the impact of man on nature.

***Atom*:** The smallest component of an element having the chemical properties of the element; the atom consists of a nucleus containing combinations of neutrons and protons and one or more electrons bound to the nucleus by electrical attraction.

***Atmospheric circulation*:** The motion within the atmosphere that results from inequalities in pressure over the earth's surface. When the average of the entire globe is considered, it is referred to as the general circulation of the atmosphere.

***Biodiversity*:** The type and variety of living organisms; biological diversity.

***Birth rate*:** The ratio between births and individuals in a specified population and time, commonly expressed by the number of live births per hundred or per thousand per year.

***Carrying capacity*:** The population of organisms that an area will support without undergoing deterioration.

***CFCs*:** Chlorofluorocarbons.

***Chlorofluorocarbons (CFCs)*:** The synthetic organic compounds composed of methyl ($CH^3$) groups, chlorine, and fluorine.

***CITES*:** Convention on International Trade in Endangered Species.

***Climate*:** All of the types of weather that occur at a given place over time.

***Climatic optimum*:** A period around 5500 years ago when the climate of Europe was warmer than today.

***Climatic regime*:** The annual cycles associated with various climatic elements; for example, the thermal regime is the seasonal patterns of temperature, and the moisture regime is the seasonal pattern of precipitation.

***Closed system*:** A system with no exchange of matter or energy through the boundaries.

***Compound*:** A pure substance composed of two or more elements whose composition is constant, such as water ($H^2O$) and common table salt (NaCl).

***Core*:** The central portion of Earth enriched in iron and nickel and having a radius of about 3379 km. The innermost portion of the core is in a molten state and the outer is solid.

***CRP*:** Conservation Reserve Program.

***Crust*:** The outer layer of Earth enriched in silicon, sodium, and potassium and having a thickness of about 35 km beneath the continents and 10 km under the oceans.

***Deflation*:** The lifting and removal of earth particles by wind action.

***Dendrochronology*:** The analysis of the annual growth rings of trees, leading to the determination of climatic conditions in the past.

***Desertification*:** A process of environmental degradation that results in sterile land.

***Diurnal*:** Occurring in the daytime or having a daily cycle.

***Dobson Unit*:** One Dobson unit is equal to a layer of ozone 1/100 mm thick at 0°C and one atmosphere of pressure. Mean ozone thickness is 3.2 mm or 320 Dobson units.

***Electromagnetic spectrum*:** The range of energy that is transferred as wave motions and that does not require any intervening matter to make the transfer. The waves travel at the speed of light, 186,000 mi/s.

***Electromagnetic waves*:** Waves characterized by variations of electric and magnetic fields.

***Environment*:** Total surroundings, habitat.

***EPA*:** Environmental Protection Agency.

***Equilibrium*:** In a state of balance.

***Erosion*:** The set of processes by which the surface of Earth is worn away by the action of water, wind, glacial ice, etc.

***Fauna*:** The assemblage of animals found in a given area.

***GCM*:** Global circulation model.

***Greenhouse effect*:** The process by which the heating of the atmosphere is compared to a common greenhouse. Sunlight (shortwave radiation) passes through the atmosphere to reach the earth. The energy reradiated by the earth is at a longer wavelength, and its return to space is inhibited by atmospheric carbon dioxide and water vapor. This process acts to increase the temperature of the lower atmosphere.

***Greenhouse gas*:** Atmospheric gases such as water vapor, carbon dioxide, methane, and nitrous oxide that are capable of absorbing radiant energy from Earth's surface.

***Groundwater*:** Water found in voids and interstices below the water table. It is water in the saturated zone.

***GTC*:** 1 billion tons of carbon.

***Hadley cell*:** A convectional cell operating as part of the general circulation located approximately between the Tropic of Cancer or the Tropic of Capricorn and the equator.

***Hydrocarbon*:** A compound made up of the elements carbon and hydrogen.

***Ice Age*:** A glacial epoch, especially the last Pleistocene Epoch beginning about 1.8 million years ago.

***IPCC*:** Inter-agency Panel of Climatic Change.

***Isotope*:** Any of two or more forms of a chemical element having the same atomic number, that is the same number of protons in the nucleus, but having different numbers of neutrons in the nucleus, therefore having different atomic weights. There are 275 isotopes of the stable elements and more than 800 radioactive isotopes. Every element has known isotopic forms.

***Infrared radiation*:** Radiation in the range longer than red. Most sensible heat radiated by the earth and other terrestrial objects is in the form of infrared waves.

***Ionosphere*:** A zone of the upper atmosphere characterized by gases that have been ionized by solar radiation.

***Latent energy*:** Energy temporarily stored or concealed, such as the heat contained in water vapor.

***Lunar*:** Pertaining to the moon.

***Jet stream*:** A high-speed flow of air that occurs in narrow bands of the upper air westerlies.

***Meridional flow*:** The movement of air in a north-south direction or along a meridian.

***NOAA*:** National Oceanographic and Atmospheric Administration.

***Open system*:** A system in which energy or matter, or both, flow in and out.

***Ozone*:** Oxygen in the triatomic form ($O_3$); highly corrosive and poisonous.

***Ozone layer*:** The layer of ozone, 25 km above the earth's surface, that absorbs ultraviolet radiation from the sun.

***PCBs*:** Polychlorinated Biphenyls.

***Perennial*:** A plant that grows all year. A perennial stream flows all the time except during severe drought.

***Periodic*:** Occurring or appearing at regular intervals, at the sun's rising and setting.

***Pesticide*:** Any one of a large group of chemicals used to control insect populations.

***pH*:** The negative logarithm of the effective hydrogen ion concentration or hydrogen ion activity used in expressing both acidity and alkalinity on a scale whose values run from 0 to 14 with 7 representing neutrality; numbers less than 7 denote increasing acidity, and numbers greater than 7 increasing alkaline conditions.

***Photochemical*:** A chemical change that either releases or absorbs radiation.

***Photoperiod*:** The period of each day when direct solar radiation reaches the earth's surface, approximately sunrise to sunset.

***Planetesimal*:** One of the small celestial bodies that fused together to form the planets of the solar system.

***Plate tectonics*:** The theory of global tectonics in which the lithosphere is divided into a number of crustal plates that move on the underlying plastic asthenosphere. These plates may collide with, slide under, or move past adjacent plates.

***ppb*:** Parts per billion.

***ppbv*:** Parts per billion by volume,

***ppm*:** Parts per million.

***ppmv*:** Parts per million by volume.

***Radiation*:** Energy transfer from the sun to Earth.

***Recurrence interval*:** The average time span between events of a given magnitude.

***Saltation*:** The movement of rock particles by water or wind by bouncing them along the surface.

***Savanna*:** A tropical grassland.

***Solar constant*:** The mean rate at which solar radiation reaches the earth.

***Solar radiation*:** Electromagnetic energy emitted by the sun.

***Stratosphere*:** The atmosphere between the troposphere and mesosphere; also the main site for ozone formation.

***Stromatolite*:** A fossil laminated structure usually found in limestones and constructed by algae and bacteria. The structure may have a rounded, columnar, or more complex form.

***Steady State*:** An open system in which the external and internal relationships are such that there is internal balance or equilibrium.

***Subsystem*:** A part of a larger system which can itself be treated as a system.

***System*:** A series of objects or entities which are linked together and bounded, and which interact for a common purpose or purposes.

***Temporal*:** Having to do with time; taking place through time.

***Terrestrial*:** Pertaining to the land, as distinguished from the sea or air.

***Tropopause*:** The upper boundary zone of the troposphere, marked by a discontinuity of temperature and moisture.

***Troposphere*:** The lower layer of the atmosphere marked by decreasing temperature, pressure, and moisture with height; the layer in which most day-to-day weather changes occur.

***Ultraviolet radiation*:** Radiation of a wavelength shorter than violet. The invisible ultraviolet radiation is largely responsible for sunburn.

***Volcanism*:** All forms of volcanic activity.

***Weather*:** The state of the atmosphere at any one point in time and space.

***Zonal flow*:** Flowing along the parallels of latitude, or east and west.

# appendix

# Declaration on the Human Environment

The United Nations Conference on the Human Environment, having met at Stockholm from 5 to 16 June 1972, having considered the need for a common outlook and for common principles to inspire and guide the peoples of the world in the preservation and enhancement of the human environment,

Proclaims that:

1. Man is both creature and molder of his environment, which gives him physical sustenance and affords him the opportunity for intellectual, moral, social and spiritual growth. In the long and tortuous evolution of the human race on this planet a stage has been reached when, through the rapid acceleration of science and technology, man has acquired the power to transform his environment in countless ways and on an unprecedented scale. Both aspects of man's environment, the natural and the man-made, are essential to his well-being and to the enjoyment of basic human rights—even the right to life itself.

2. The protection and improvement of the human environment is a major issue which affects the well-being of peoples and economic development throughout the world; it is the urgent desire of the peoples of the whole world and the duty of all Governments.

3. Man has constantly to sum up experience and go on discovering, inventing, creating and advancing. In our time, man's capability to transform his surroundings, if used wisely, can bring to all peoples the benefits of development and the opportunity to enhance the quality of life. Wrongly or heedlessly applied, the same power can do incalculable harm to human beings and

the human environment. We see around us growing evidence of man-made harm in many regions of the earth: dangerous levels of pollution in water, air, earth and living beings; major and undesirable disturbances to the ecological balance of the biosphere; destruction and depletion of irreplaceable resources; and gross deficiencies harmful to the physical, mental and social health of man, in the man-made environment, particularly in the living and working environment.

4. In the developing countries most of the environmental problems are caused by underdevelopment. Millions continue to live far below the minimum levels required for a decent human existence, deprived of adequate food and clothing, shelter and education, health and sanitation. Therefore, the developing countries must direct their efforts to development, bearing in mind their priorities and the need to safeguard and improve the environment. For the same purpose, the industrialized countries should make efforts to reduce the gap between themselves and the developing countries. In the industrialized countries, environmental problems are generally related to industrialization and technological development.

5. The natural growth of population continuously presents problems for the preservation of the environment, and adequate policies and measures should be adopted, as appropriate, to face these problems. Of all things in the world, people are the most precious. It is the people that propel social progress, create social wealth, develop science and technology and, through their hard work, continuously transform the human environment. Along with social progress and the advance of production, science and technology, the capability of man to improve the environment increases with each passing day.

6. A point has been reached in history when we must shape our actions throughout the world with a more prudent care for their environment consequences. Through ignorance or indifference we can do massive and irreversible harm to the earthly environment on which our life and well-being depend. Conversely, through fuller knowledge and wiser action, we can achieve for ourselves and our posterity a better life in an environment more in keeping with human needs and hopes. There are broad vistas for the enhancement of environmental quality and the creation of a good life. What is needed is an enthusiastic but calm state of mind and intense but orderly work. For the purpose of attaining freedom in the world of nature, man must use knowledge to build in collaboration with nature a better environment. To defend and improve the human environment for present and future generations has become an imperative goal for mankind—a goal to be pursued together with, and in harmony with, the established and fundamental goals of peace and of world-wide economic and social development.

7. To achieve this environmental goal will demand the acceptance of responsibility by citizens and communities and by enterprises and institutions at every level, all sharing equitably in common efforts. Individuals in all walks of life as well as organizations in many fields, by their values and the sum of their actions, will shape the world environment of the future. Local and national governments will bear the greatest burden for large-scale environmental policy and action within their jurisdictions. International co-operation is also needed in order to raise resources to support the developing countries in carrying out their responsibilities in this field. A growing class of environmental problems, because they are regional or global in extent or because they affect the common international realm, will require extensive co-operation among nations and action by international organizations in the common interest. The Conference calls upon Governments and peoples to exert common efforts for the preservation and improvement of the human environment, for the benefit of all the people and for their posterity.

## DECLARATION OF PRINCIPLES

States the common conviction that

*Principle 1.* Man has the fundamental right to freedom, equality and adequate conditions of life, in an environment of a quality that permits a life of dignity and well-being, and he bears a solemn responsibility to protect and improve the environment for present and future generations. In this respect, policies promoting or perpetuating *apartheid* racial segregation, discrimination, colonial and other forms of oppression and foreign domination stand condemned and must be eliminated.

*Principle 2.* The natural resources of the earth including the air, water, land, flora and fauna and especially representative samples of natural ecosystems must be safeguarded for the benefit of present and future generations through careful planning or management, as appropriate.

*Principle 3.* The capacity of the earth to produce vital renewable resources must be maintained and, wherever practicable, restored or improved.

*Principle 4.* Man has a special responsibility to safeguard and wisely manage the heritage of wildlife and its habitat which are now gravely imperiled by a combination of adverse factors. Nature conservation including wildlife must therefore receive importance in planning for economic development.

*Principle 5.* The non-renewable resources of the earth must be employed in such a way as to guard against the danger of their future exhaustion and to ensure that benefits from such employment are shared by all mankind.

*Principle 6.* The discharge of toxic substances or of other substances and the release of heat, in such quantities or concentrations as to exceed the capacity of the environment to render them harmless, must be halted in order to ensure that serious or irreversible damage is not inflicted upon ecosystems. The just struggle of the peoples of all countries against pollution should be supported.

*Principle 7.* States shall take all possible steps to prevent pollution of the seas by substances that are liable to create hazards to human health, to harm living resources and marine life, to damage amenities or to interfere with other legitimate uses of the sea.

*Principle 8.* Economic and social development is essential for ensuring a favorable living and working environment for man and for creating conditions on earth that are necessary for the improvement of the quality of life.

*Principle 9.* Environmental deficiencies generated by the conditions of underdevelopment and natural disasters pose grave problems and can best be remedied by accelerated development through the transfer of substantial quantities of financial and technological assistance as a supplement to the domestic effort of the developing countries and such timely assistance as may be required.

*Principle 10.* For the developing countries, stability of prices and adequate earnings for primary commodities and raw material are essential to environment management since economic factors as well as ecological processes must be taken into account.

*Principle 11.* The environmental policies of all States should enhance and not adversely affect the present or future development potential of developing countries, nor should they hamper the attainment of better living conditions for all, and appropriate steps should be taken by States and international organizations with a view to reaching agreement on meeting the possible national and international economic consequences resulting from the application of environmental measures.

*Principle 12.* Resources should be made available to preserve and improve the environment, taking into account the circumstances and particular requirements of developing countries and any costs which may emanate from their incorporating environmental safeguards into their development planning and the need for making available to them, upon their request, additional international technical and financial assistance for this purpose.

*Principle 13.* In order to achieve a more rational management of resources and thus to improve the environment, States should adopt an integrated and co-ordinated approach to their development planning so as to ensure that development is compatible with the need to protect and improve the human environment for the benefit of their population.

*Principle 14.* Rational planning constitutes an essential tool for reconciling any conflict between the needs of development and the need to protect and improve the environment.

*Principle 15.* Planning must be applied to human settlements and urbanization with a view to avoiding adverse effects on the environment and obtaining maximum social, economic and environmental benefits for all. In this respect projects which are designed for colonialist and racist domination must be abandoned.

*Principle 16.* Demographic policies, which are without prejudice to basic human rights and which are deemed appropriate by Governments concerned, should be applied in those regions where the rate of population growth or excessive population concentrations are likely to have adverse effects on the environment or development, or where low population density may prevent improvement of the human environment and impede development.

*Principle 17.* Appropriate national institutions must be entrusted with the task of planning, managing or controlling the environmental resources of States with the view to enhancing environmental quality.

*Principle 18.* Science and technology, as part of their contribution to economic and social development, must be applied to the identification, avoidance and control of environmental risks and the solution of environmental problems and for the common good of mankind.

*Principle 19.* Education in environmental matters, for the younger generation as well as adults, giving due consideration to the under-privileged, is essential in order to broaden the basis for an enlightened opinion and responsible conduct by individuals, enterprises and communities in protecting and improving the environment in its full human dimension. It is also essential that mass media of communications avoid contributing to the deterioration of the environment, but, on the contrary, disseminate information of an educational nature, on the need to protect and improve the environment in order to enable man to develop in every respect.

*Principle 20.* Scientific research and development in the context of environmental problems, both national and multinational, must be promoted in all countries, especially the developing countries. In this connection, the free flow of up-to-date scientific information and transfer of experience must be supported and assisted, to facilitate the solution of environmental problems; environmental technologies should be made available to developing countries on terms which would encourage their wide dissemination without constituting an economic burden on the developing countries.

*Principle 21.* States have, in accordance with the Charter of the United Nations and the principles of international law, the sovereign right to exploit their own resources pursuant to their own environmental policies, and the responsibility to ensure that activities within their jurisdiction or control do not cause damage to the environment of other States or of areas beyond the limits of national jurisdiction.

*Principle 22.* States shall co-operate to develop further the international law regarding liability and compensation for the victims of pollution and other environmental damage caused by activities within the jurisdiction or control of such States to areas beyond their jurisdiction.

*Principle 23.* Without prejudice to such criteria as may be agreed upon by the international community, or to standards which will have to be determined nationally, it will be essential in all cases to consider the systems of values prevailing in each country, and the extent of the applicability of standards which are valid for the most advanced countries but which may be inappropriate and of unwarranted social cost for the developing countries.

*Principle 24.* International matters concerning the protection and improvement of the environment should be handled in a co-operative spirit by all countries, big or small, on an equal footing. Co-operation through multilateral or bi-lateral arrangements or other appropriate means is essential to effectively control, prevent, reduce and eliminate adverse environmental effects resulting from activities conducted in all spheres, in such a way that due account is taken of the sovereignty and interest of all States.

*Principle 25.* States shall ensure that international organizations play a co-ordinated, efficient and dynamic role for the protection and improvement of the environment.

*Principle 26.* Man and his environment must be spared the effects of nuclear weapons and all other means of mass destruction. States must strive to reach prompt agreement, in the relevant international organs, on the elimination and complete destruction of such weapons.

# Index

### S



### T



### U



### V



### W